Reconnaissance Geology of the State of Baja California

The Geological Society of America, Inc.
Memoir 140

Reconnaissance Geology of the State of Baja California

R. GORDON GASTIL
Department of Geology, San Diego State University
San Diego, California 92182

RICHARD P. PHILLIPS
Environmental Studies Program, University of San Diego
Alcala Park, San Diego, California 92110

EDWIN C. ALLISON
Department of Geology, San Diego State University
San Diego, California 92182

1975

Copyright 1975 by The Geological Society of America, Inc.
Library of Congress Catalog Card Number 74-83806
I.S.B.N. 0-8137-1140-1

550.62
G345m
no. 140
1975

Published by
THE GEOLOGICAL SOCIETY OF AMERICA, INC.
3300 Penrose Place
Boulder, Colorado 80301

Printed in The United States of America

To the memory of

CHARLES VICTOR ANTHENIL
*who died from a fall June 18, 1966,
while mapping in the state of Baja California*

and

EDWIN CHESTER ALLISON
*who died on January 1, 1971,
while mapping in the state of Sonora*

*The Memoir series was originally made possible
through the bequest of
Richard Alexander Fullerton Penrose, Jr.*

Contents

Acknowledgments		xiii
Abstract		1
1	Introduction	3
	Maps and aerial photographs	3
	Previous geologic investigations	5
	Outline of geologic history	7
	Geologic map of Baja California	8
	Geographic names	8
2	Prebatholithic terrane	11
	"Franciscan" belt	11
	Volcanic-volcaniclastic belt	13
	Volcanic dikes	20
	Shale-sandstone belt	21
	Paleozoic metasedimentary belt	23
	Problems of nomenclature	24
3	Mesozoic thermal event	27
	Metamorphic framework	27
	Peninsular Ranges batholith	29
	Mineral assemblages	30
	Chemical analysis of batholithic rocks	33
	Sizes of plutons	36
	Ages of plutons	39
	Ages of metamorphism and cooling	40
4	Late Cretaceous and early Tertiary time	43
	Postbatholithic Cretaceous rocks	43
	Stratigraphic nomenclature	43
	Sedimentary rocks	45
	Lower Tertiary rocks	48
	Stratigraphic nomenclature	48
	Sedimentary rocks	49
	Volcanic rocks	57

5 Mid-Cenozoic volcanism ... 59
Stratigraphic nomenclature ... 59
Volcanic rocks ... 62
    Previous observations ... 62
    Terminology ... 62
    K-Ar mineral ages ... 63
    Chemical analyses ... 63
    Gulf of California volcanic province ... 66
    Pacific coast volcanic province ... 70

6 Flooding of the Gulf of California and the continental borderland ... 75
Pliocene Epoch on the Pacific coast ... 75
Earliest marine strata in the Gulf of California depression ... 76
Pliocene rocks of the Gulf of California depression ... 76
Peninsular interior during late Cenozoic time ... 79
Conclusions ... 79

7 Pleistocene and Holocene Epochs ... 81
Pacific coastal terraces ... 81
Nonmarine coastal deposits ... 83
Interior deposits ... 83
Gulf of California coastal deposits ... 84
Volcanism ... 84
Submerged portion of the continental borderland ... 86
Pacific coastal lagoons ... 87
Gulf of California ... 87

8 Geomorphology ... 89
Previous investigations ... 89
Geomorphic history ... 90
Pacific coastal provinces ... 92
Coastal mountain provinces ... 93
Northern highland plateau provinces ... 94
Highland valley provinces ... 97
Southern highland plateau provinces ... 98
Gulf of California provinces ... 102

9 Geophysics ... 107
Heat-flow surveys ... 107
Magnetic surveys ... 108
Earthquake seismicity ... 110
Seismic surveys ... 113
Gravity surveys ... 116

10 Structure ... 121
Prebatholithic and batholithic structures ... 121
Upper Cretaceous structures ... 123
Major structural lineaments ... 124
Structural provinces ... 127
    Continental borderland ... 127
    Stable peninsula ... 128
    Gulf of California depression ... 131

CONTENTS ix

11 Regional tectonic pattern 137
    Mesozoic arc 137
    Tectonic framework 138
    Hypotheses for the origin of the Gulf of California 139
    Discussion of recent hypotheses 141

12 Economic geology 145
    History of mineral exploration and exploitation 145
    Mineral provinces 147
        Volcanic copper-iron province 148
        Schist-gold province 149
        Scheelite deposits of the carbonate province 149
        Cenozoic hydrothermal province 150
        Placer gold province 153
    Other resources 153

References cited 157

APPENDIXES

1. Master's theses 154
2. Senior and other undergraduate research papers 155

(On microfiche in pocket)

3. Mineralogy and petrography of metavolcanic rocks, Tijuana-Tecate area
4. Location and petrographic description of chemically analyzed plutonic rocks
5. Variation diagrams for chemically analyzed plutonic rocks
6. Citations of Neogene marine strata near the center of the state of Baja California
7. Locations and petrographic descriptions of volcanic rocks
8. Original citation of evidence for deep recent submergence of the peninsula
9. Gravity base stations
10. Analysis of iron ore from San Vicente area
11. Analysis of hot springs in Arroyo Volcán
12. Analyses of brines from Cerro Prieto geothermal well M-3
13. Chemical analyses of brines from Salton Sea area
14. Comparison of relative abundances of elements in Boleo copper ore with relative abundances in igneous rocks

Note: Hard copy may be ordered from the Documents Secretary, Geological Society of America, Boulder, Colorado 80301. Refer to appendix number and GSA-SM-75-10.

FIGURES

1. Regional setting 4
2. Index to map sources 6
3. Index to geologic map reliability 9
4. Distribution of prebatholithic rocks 12
5. Stratigraphic section of the type Alisitos Formation 15
6. Stratigraphic section of the Alisitos Formation, east of El Rosario 16
7. Anticline in prebatholithic strata 17
8. Stratigraphic section of pre-Alisitos strata, Arroyo San José 20
9. Stratigraphic section, southeast of El Burro 23
10. Stratigraphic section of metamorphic rocks, northern Sierra Pinta 24
11. Northern end of Sierra Pinta 25
12. Paleozoic strata, Arroyo Volcán 26
13. Metamorphic-tectonic zones 28
14. El Pinal pluton 31
15. Quartz-potassium feldspar-plagioclase diagram 32
16. Relation of $K_2O + Na_2O$ indexes to distance from axis of the Peninsular Ranges batholith 37
17. Size distribution of Baja California plutons compared with those of the Sierra Nevada 38
18. Shapes of some concentrically structured plutons 39
19. Relation of dike swarms to plutons near Vallecitos 40
20. Distribution of Upper Cretaceous and lower Tertiary rocks 42
21. Correlation of postbatholithic Cretaceous, Paleocene, and Eocene formations . . 45
22. Diagrammatic section through Mesa San Carlos 52
23. Locations of early Tertiary stream channels 56
24. Distribution of Cenozoic volcanic rocks 58
25. Correlation of middle to upper Cenozoic formations 60
26. Histogram of ages of Cenozoic volcanic rocks 63
27. $(K_2O + Na_2O)/SiO_2$ variation diagram for Cenozoic volcanic rocks 65
28. Comparison of $(K_2O + Na_2O)/SiO_2$ ratios in Cenozoic volcanic and Mesozoic plutonic rocks 67
29. Matomí volcanic plateau 68
30. Stratigraphic section of volcanic rocks in the northern Sierra Pinta 71
31. Correlation of La Misión and Rosarito Beach sections 72
32. Distribution of Pliocene and Miocene sedimentary rocks 74
33. Marine terraces on the Gulf of California 78
34. Distribution of Quaternary sedimentary rocks 80
35. Diagrammatic sketch of terraces west of Mesa San Carlos 82
36. Isla San Luis 85
37. Pleistocene basalt flows 86
38. Geomorphic provinces 88
39. Topographic profiles across Baja California and Sonora 91
40. Suggested relation between the old erosion surface and the coastal highlands . . 94
41. North end of the Sierra Tinaja and the scarp of the Sierra Juárez 96
42. Diagrammatic sketch to illustrate the tilting of the peninsula 98
43. Exhumed erosion surface 100
44. Typical bedrock surface 101
45. South end of the Sierra Tinaja 104
46. Diagrammatic cross section of the Sierra Santa Rosa 104
47. Intersection of Gonzaga lineament with main gulf escarpment 105
48. Earthquake epicenters 111
49. Recent fault scarps 112

50. Close-up of fault scarp 113
51. Seismic refraction profile from San Clemente Island to Corona, California . . . 114
52. Seismic refraction profile across the continental borderland through
    Isla San Benito 115
53. Seismic refraction profile of the Gulf of California 115
54. Gravity base stations, traverses, and level lines 117
55. Structural provinces 122
56. Major structural elements of the continental borderland 123
57. Medial fault in the Valle San Pedro 125
58. Agua Blanca–Santo Tomás fault system 125
59. Relation of offshore magnetic anomalies to Santillán y Barrera line between
    Socorro and Punta Canoas 129
60. Hypothetical cross section across the gulf coastline north of San Felipe 132
61. Diagrammatic sketch of a fault set, north of Rancho Parral 133
62. Generalized sketch of antithetic fault sets east of the stable peninsula 134
63. Structural framework of the Californias 136
64. Interpretations of the early opening of the Gulf of California 139
65. The East Pacific Rise at the mouth of the Gulf of California 143
66. Mineral deposits in the state of Baja California 144

TABLES

1. Chemical analyses of prebatholithic volcanic rocks 15
2. Thin-section identification of granitic rocks 32
3. Composition of plutons 32
4. Chemical analyses of plutonic rocks 34
5. Modes and norms for plutonic rocks 35
6. Comparison of point-count modes and norms for the felsic components of
   plutonic rocks 36
7. K-Ar "ages" for batholithic and prebatholithic rocks 41
8. Percent clast composition of early Tertiary gravel units by area 56
9. K-Ar "ages" for volcanic rocks 57
10. Chemical analysis of Tertiary volcanic rocks 64
11. Comparison of four Miocene basalts of Baja California with Pliocene-
    Pleistocene basalts of Oregon and Idaho 67
12. Strontium isotope ratios for volcanic rocks 67
13. Cobble counts in the Mira al Mar Member of the Rosarito Beach Formation . . 71
14. Modes of basalt from the Rosarito Beach Formation 72

PLATES

1. Reconnaissance geologic map
   (A three-sheet map, prepared to accompany this Memoir, is included in plastic
   wrap.)
   Addendum to Plate 1 xii
2. Plutonic rocks . . . . in pocket
3. Fault map . . . . in pocket
4. Topographic profiles . . . . in pocket
5. Bouguer gravity map . . . . in pocket
6. Geologic cross sections . . . . in pocket

ADDENDUM TO PLATE 1

In the area east of Tijuana and north of Rodriquez, Tpm has been used to include all post-Eocene strata (as in Hertlein and Grant, 1939). After our mapping, the lower portion of this strata was named the Otay Formation (Artim and Pinckney, 1973) and correlated with the Rosarito Beach Formation (Minch, 1967) that crops out south of Tijuana.

Islas los Coronados, just west of Tijuana, are not included on the map. They have been mapped in detail by Tom Lamb (unpub. data).

Isla Cedros, west of Guerrero Negro, has been mapped by Frank Kilmer, Humboldt State University, Arcata, California (unpub. data).

Valle de las Palmas, southeast of Tijuana, has been mapped by Martin Frazer (1972, unpub. data). His revised map shows two ages of older conglomerate (mapped as Redonda and Buenos Aires Formations after Flynn, 1970), another small body of Tertiary andesite like those east of Tijuana, and additional northwest-trending faults.

The granodiorite body shown east of Punta China is actually gabbro. Isla San Geronimo, south of Punta Baja, is mistakenly marked pb?. It should be included in the Rosario Group.

The observatory on top of the Sierra San Pedro Mártir is now reached by following the road east from Rancho San José.

Laguna Amaya and Agua Amaya indicated northwest of Bahía de los Angeles should be Laguna Amarga and Agua Amarga, respectively.

The region between Rancho Grande and El Arco has been remapped by David Barthelmy (unpub. data).

Pliocene marine strata were omitted from Isla Salsipuedes and Isla San Lorenzo. The chain of small islands San Lorenzo to Lobos and, farther north, the chain of islands El Muerto to San Luis have been mapped in detail by Robert Rossetter (App. 1).

Acknowledgments

The work presented here began in 1963. It was performed as a cooperative project by students and staff of San Diego State University, the Escuela Superior de Ciencias Marinas of the Universidad Autónoma de Baja California at Ensenada, the Instituto de Geología de México, and the San Diego Museum of Natural History. It was supported in part by grants from the National Science Foundation. Faculty members who participated were Edwin C. Allison, R. Gordon Gastil, Richard P. Phillips, Ellis E. Roberts, Gary L. Peterson, and Daniel Krummenacher from San Diego, and César Obregón and Rudolfo Malpica from Ensenada. Johathan Bushee, John Minch, and Donald Fife served as project supervisors on particular problems.

Students who contributed to the project include the following National Science Foundation Undergraduate Research participants: Roberto Aiguere, Robert Andersen, José Arciniego, Alfredo Barragón, Linda Beltz, Horacio Benites, Phillip Birkhahn, Francis Blake, Marilyn Bourgeoise, John Brown, Robert Burk, Lucy Chamlee, Bruce Clardy, Rodolfo Cruz, Carlos de Alba, Juley Dwyer, William Elliott, William Estavillo, James Evans, Martin Fitzurka, Michael Flynn, John Garcia, Judith Gassaway, Harriet Goldis, William Grey, Natalie Harper, Herbert Harris, Arthur Henry, Ramón Hernández, David Hicks, George R. Hofman, Deborah Hughes, Alton James, Wallace Jensky, Paul Johansing, William Jones, John Kaiser, Ronald LaBorde, Raymundo Lecuanda, Stephen Leedom, Richard Lehtola, Edward Lubitz, Robert Lunceford, John Marienthal, Claude Marshall, David McGee, John Minch, George Morgan, J. R. Morgan, David Murray, Katsuo Nishikawa, Carlos Perez, Bruce Peterson, Michael Ploessil, Charles Preston, Arthur Ravenscroft, Mark Rogers, Robert Rossetter, Kenneth Schulte, Richard Sherer, Patricia Sivils, William Skinner, Robert Slyker, Warren Smith, Henry Snyder, Michael Sommer, David Strong, Charles Stuart, Walter Timm, Wilbert Ulman, Francisco Valdez, Francisco Vidal, Gary Vogt, Charles Weisenburg, Clarence Wendt, and Bryan Worthington.

Other students who contributed were Randal Ashley, Robert Bell, William Carpenter, Ali Diaeldin, Barbara Geyer, Peter Helander, Peter Hord, Joseph Itson, Terry Kebort, Lloyd Lehrer, Lawrence Pendarvis, William Pfister, Dale Stickney, James Stroh, Margaret Stroh, and John Turner.

Master's degree theses by students listed in Appendix 1 constitute important contributions.

Leon Silver of California Institute of Technology, William Morris of Occidental College, and Avaril Cross of Michigan State University cooperated during field studies; Wendell Duffield and Raymond Elliot of Stanford University and the U.S. Geological Survey and Fred Barnard, then of the University of Colorado, contributed their mapping.

The final drafting of the figures was done by Paula Yeager.

Abstract

The prebatholithic terrane of Baja California is represented by a western belt of rocks like those of the Franciscan assemblage, a belt of Mesozoic volcanic-volcaniclastic rocks, a belt of Mesozoic(?) slate and metasandstone, and, along the Gulf of California, Paleozoic rocks of various types. The volcanic-volcaniclastic rocks are, in part, contemporaneous with batholithic emplacement.

The Mesozoic thermal event that produced regional metamorphism and the emplacement of batholithic rocks can be described in terms of four metamorphic zones: (1) essentially unmetamorphosed rocks, (2) slate to phyllitic rocks, (3) schist and amphibolite, and (4) rocks in which the premetamorphic fabric has been entirely destroyed. Most of the discrete plutons are found in zone 3; zone 4 includes generalized metamorphic-plutonic rocks that have been deep in the crust for a prolonged time. The granitic exposures are made up of tonalite (73 percent), granodiorite (23 percent), and gabbro (2 percent). The plutons of largest diameters are formed of tonalite; plutons of smallest diameters, of gabbro and granite. Gabbro is common in the western half of the batholithic belt and rare in the eastern half. No systematic increase in $(K_2O + Na_2O)/SiO_2$ is recognized across the peninsula.

The postbatholithic interval began about 90 m.y. ago. The erosion of the western portion of the metamorphic-plutonic terrane was accomplished within the first 20 m.y. Potassium-argon dating indicates that the basement rock underlying the gulf area cooled somewhat later. Upper Cretaceous deposits were derived from the erosion of the western part of the peninsula. In contrast, the Paleocene-Eocene deposits included detritus derived from areas now within the Gulf of California depression and from areas farther east. During the Eocene Epoch, the northern part of the peninsula was a pediplain not far above sea level.

Although volcanism occurred locally during early Cenozoic time, widespread calc-alkalic volcanism and the downfaulting of the gulf area began during late Oligocene time. This volcanic phase was followed about 9 m.y. ago by the entry of the sea into the northern part of the gulf. About 6 to 8 m.y. ago, the gulf area was subjected to folding, uplift, and erosion that removed the early Cenozoic volcanic rocks from wide areas. During this interval, the peninsula also was uplifted and stripped of much of its Cenozoic cover, and the early Cenozoic pediplain and Late Cretaceous coastline were exhumed. Alkalic basalt and basaltic andesite then poured across the dissected Pliocene-Pleistocene surface.

The topography of the peninsula displays remnants of both the erosional surfaces and coastal configurations that were produced during Late Cretaceous and early

Abstract

Cenozoic time. They were disrupted by volcanism and faulting in Miocene and later time, uplifted in the Pliocene Epoch, and deeply dissected by subsequent erosion.

Structurally, Baja California is divided into the stable western and central peninsula, the unstable continental borderland to the west, and the gulf depression to the east. The borderland has been structurally distinct since at least Late Cretaceous time. The gulf depression has undergone about 50 percent dilation since the Oligocene Epoch, and extensional deformation is probably continuing.

A regional gravity survey of the state of Baja California shows the peninsula to be generally in isostatic adjustment. The Bouguer anomaly map reveals a gravity high that extends down the west coast from the Agua Blanca fault to at least lat 28° N.

Mineral exploration indicates that there are several distinct mineral provinces. Three of these are the volcanic copper-iron province, which corresponds to the Mesozoic volcanic-volcaniclastic belt; the schist-gold province, which corresponds to the Mesozoic(?) slate and metasandstone belt; and the Cenozoic hydrothermal province of the gulf area. The most fascinating and least understood of these is the Cenozoic hydrothermal province, where both base and precious metals are probably being deposited today.

Key words: areal geology, economic geology, geochemistry, geochronology, geomorphology, igneous and metamorphic petrology, solid-earth geophysics, stratigraphic and historical geology, structural geology, absolute age, batholiths, bibliography, bivalvia, Cenozoic, chemical analysis, continental shelf, craters, Cretaceous, deformation, deserts, diapirs, diastrophism, dikes, drainage changes, earthquakes, Eocene, erosion surfaces, evaporites, extrusive rocks, faults, folds, foliation, geophysical surveys, geophysics, geosynclines, geothermal energy, historical geology, igneous rocks, intrusions, Jurassic, maps, Mesozoic, metamorphic rocks, mineral waters, Miocene, orogeny, Paleocene, paleogeography, paleontology, Paleozoic, Permian, petrology, placers, Pleistocene, Pliocene, plutons, Quaternary, rift valleys, sea-floor spreading, sedimentary rocks, springs, stratigraphy, structural analysis, structural geology, tectonics, terraces, Tertiary, thermal waters, Triassic, uplifts, volcanic rocks, volcanism, volcanoes.

1
Introduction

Politically, the peninsula of Baja California is divided into the northern Estado (state) de Baja California and the southern Territorio (territory) de Baja California. The border between the two corresponds to lat 28° N. The state extends northwestward from lat 28° N. approximately 550 km to lat 33° N. and varies in width from 200 km along the international boundary with the state of California, U.S.A., to less than 100 km at San Luis Gonzaga. Physiographically, it consists of a narrow and discontinuous Pacific coastal belt, a backbone of mountains and central plateaus, and an eastern province of basins and small desert ranges. Geologically, these physiographic domains correspond to the continental borderland, the Peninsular Ranges, and the Gulf of California depression, which is a portion of the Basin and Range province (Fig. 1).

Beal's work (Marland Oil Company of Mexico, 1924) was accomplished a half century ago. In some ways, little has changed. The walls of Misión San Fernando continue to crumble, whereas the candles still flicker at Misión Santa Gertrudis. The Castros, Villavicencios, Arces, and Cotas still own their isolated ranches that were established more than a century ago, but now most of them can be reached by truck.

Since Beal's visit, Ensenada, Tijuana, Tecate, San Felipe, and Mexicali have become cities connected to the modern world by good highways and television, and now a narrow paved highway extends the length of the peninsula. But away from the pavement, the roads still change with each storm. Habitations are so far apart that each adobe (abandoned or not) constitutes a place name on the map. In the more remote areas, the sewing machine is still packed in on the back of a mule, and shoes, bedsprings, and riatas are still made from local rawhide.

MAPS AND AERIAL PHOTOGRAPHS

The largest scale maps available to the public are the 1:500,000 shaded contour maps of the Comisión Intersecretarial Coordinadora del Levantamiento de la Cartográfica de la República Mexicana (1958) and the 1:100,000 planimetric maps of the Secretaría de Recursos Hidráulicos (Tamayo, 1958). These maps appear to be the principal sources for the commonly available chart for air navigation

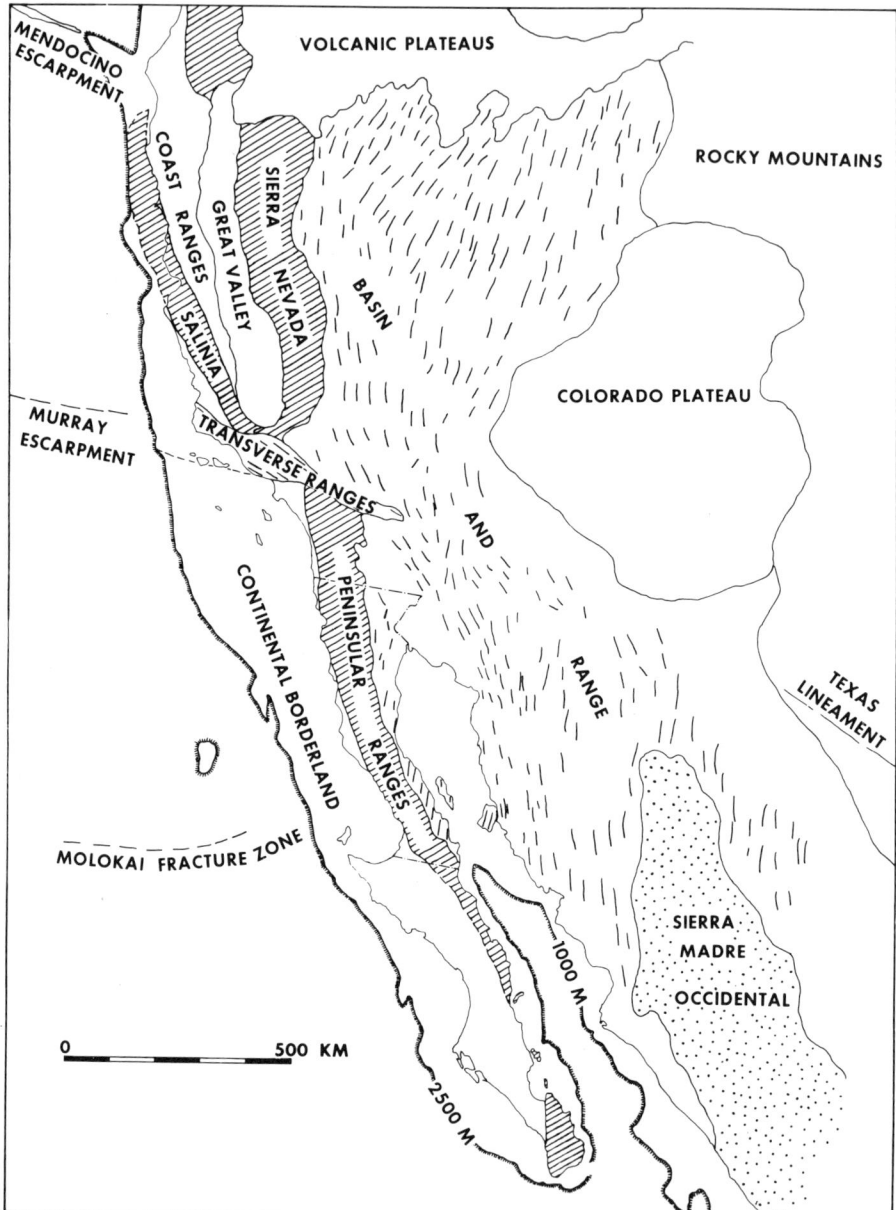

Figure 1. Regional setting of the state of Baja California. The continental edge is marked by the 1,000-m isobath in the Gulf of California and the 2,500-m isobath in the Pacific Ocean.

in Baja California (Aerospace Center of Defense Mapping Agency, 1969). Unfortunately, all three of these groups of maps are very misleading in respect to roads, place names, and (to a lesser extent) drainages. The 1:1,000,000 road map by Gerhard and Gulick (1970) is accurate. The road map and guidebooks for Baja California (Automobile Club of Southern California, 2601 South Figueroa Street, Los Angeles, 1973) are up-to-date in most areas. Areas north of lat 30° N. are shown on 1:250,000 topographic maps produced by the U.S. Army Map Service (1958-1964), in cooperation with the government of Mexico. Their distribution is restricted, and they are not generally available. These maps were produced from excellent vertical aerial photographs, and except for isolated errors in drainage and place names, they are excellent maps for the scale. The southeast corner of the state has been similarly mapped, but the drainage and culture are incorrect.

The Consejo de Recursos Naturales no Renovables (1965), in cooperation with the United Nations, published a series of topographic, planimetric, photogeologic, and aeromagnetic maps of the western part of the peninsula between Ensenada and lat 30° N. (area 51 of Fig. 2). The topographic and aeromagnetic maps extend only as far south as lat 31° N. These maps were prepared from vertical aerial photographs and published at a scale of 1:50,000. They are available from Consejo de Recursos Naturales no Renovables, México, D.F., México.

Vertical aerial photographic coverage of the state has been flown several times, but most of the results are unavailable. We eventually were able to obtain photographs covering about 80 percent of the state (Fig. 3) and to look briefly at those covering most of the remaining 20 percent. Coverage of the coastal zone was obtained from the U.S. Office of Oceanographic Research. Some photographs were lent to us by the U.S. Geological Survey and the Consejo de Recursos Naturales no Renovables. The balance was purchased from aerial photography companies in Mexico City.

Since the completion of this work, the Mexican government has created the Comisión de Estudios del Territorio Nacional to produce complete aerial photographic and topographic map coverage of the nation for the public at reasonable prices. The aerial photographs of Baja California are now available from Sección Distribución, San Antonio, Abad 124 P.B., México, D.F., México.

PREVIOUS GEOLOGIC INVESTIGATIONS

Apparently, the first geologic observations were made by Father Johann Jakob Baegert (1771). He compared the Gulf of California to the Red Sea of the Old World and speculated that Baja California was once joined to Sonora (p. 12). He observed that Baja California was composed of "wacke and other worthless rocks" (p. 19). He identified wacke as the building stone (rhyolite tuff or tuff breccia) used so successfully by the Jesuits for building missions (p. 125).

José Longinos Martínez, primarily a botanist, chemically analyzed spring waters and made notes on the rocks and minerals of the entire length of the Californias (Longinos Martínez, 1792).

The first article treating the geology of Baja California in a scientific journal was published in St. Petersburg (U.S.S.R.) in 1848 by C. Grewingk. Unfortunately, it skips from the southern half of the peninsula to Alta California (U.S.A.) and says almost nothing about our area of interest.

Modern geologic knowledge of Baja California began with William M. Gabb, who traveled the length of the peninsula in 1867. His observations from this expedition were inserted in a "Report on the Mineral Resources of the States and Territories West of the Rocky Mountains," produced by the U.S. Treasury Department under the general authorship of Browne (1868). Gabb's observations were in part rewritten and privately published by Browne in 1869. This report served as the basis of geologic knowledge of the peninsula for half a century. As an example of Gabb's

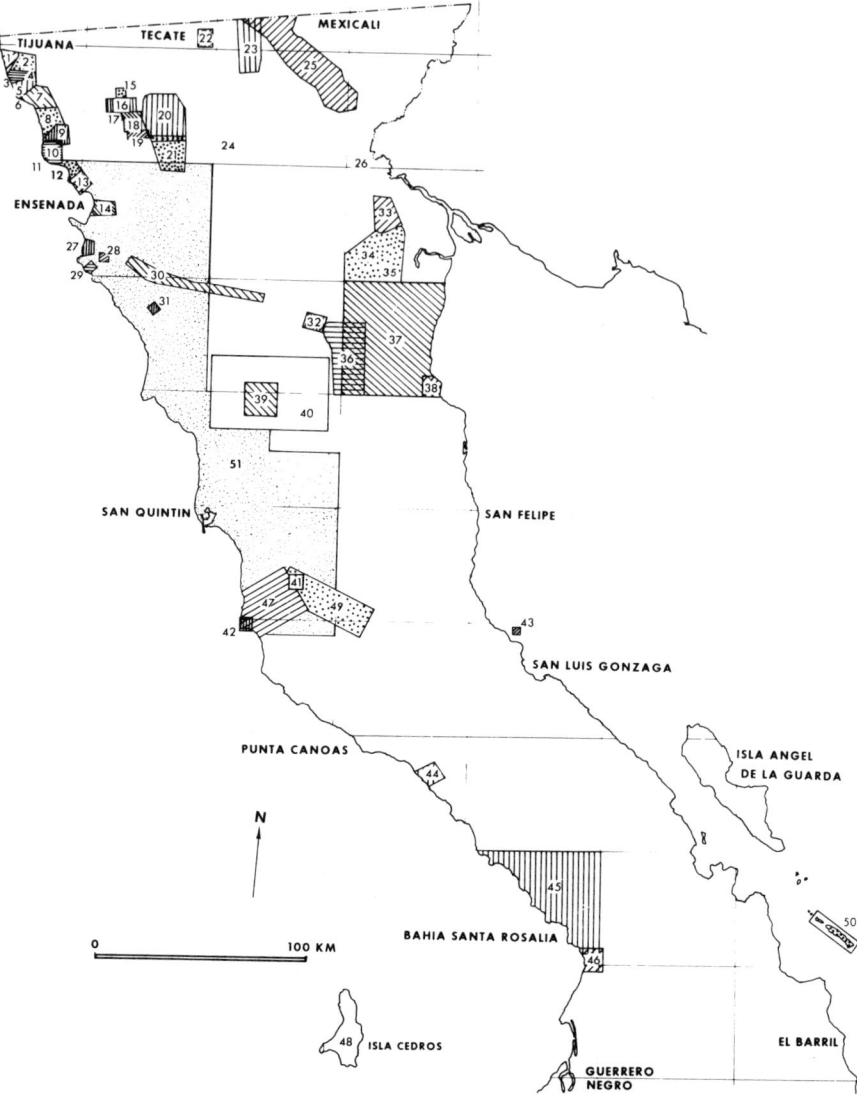

Figure 2. Index to the following map sources used for the compilation of the reconnaissance geologic map of the state of Baja California: 1, Minch (App. 1); 2, Flynn (App. 1); 3, Stuart (App. 2); 4, Kebort (App 2); 5, Turner (App. 2); 6, Hord (App. 2); 7, Worthington (App. 2), McGee (App. 1), Mickey (App. 1), Carpenter (App. 2); 8, Pendarvis (App. 2); 9, Schulte (App. 2); 10, Hofman (App. 2); 11, Itson (App. 2); 12, Bailey (App. 1); 13, Helander (App. 2); 14, Schroeder (App. 1); 15, Snyder (App. 2); 16, Bell (App. 2); 17, Peterson (App. 2); 18, Kaiser (App. 2); 19, M. Ploessil (1966, written commun.); 20, Duffield (1968); 21, R. Elliott (1967, written commun.); 22, Pfister (App. 2); 23, Advanced field class, San Diego State University (1963); 24, A. H. James (App. 2); 25, Barnard (1968b); 26, Lehtola (App. 2); 27, Acosta (App. 1); 28, Lehrer (App. 2); 29, Acosta (App. 2 [1963]); 30, Allen and others (1960); 31, Leedom (App. 2); 32, Rogers (App. 2); 33, La Borde (App. 2), McEldowney (App. 1); 34, James (App. 1); 35, Henry (App. 2); 36, Slyker (App. 1); 37, Andersen (App. 1); 38, Jones (App. 2); 39, Birkhahn (App. 2); 40, Woodford and Harriss (1938); 41, Reed (App. 1); 42, McGee (App. 2 [1965]); 43, Rossetter (App. 2); 44, Minch (1969); 45, Fife (App. 1); 46, Andersen (App. 2); 47, Kilmer (1963); 48, Kilmer (1969); 49, J. A. Noble (1957, written commun.); 50, Anderson (1950); 51, Consejo de Recursos Naturales no Renovables (1965).

insight, he (in Browne, 1869, p. 115) described the decrease in size and abundance, and the increase in rounding, of Tertiary volcanic fragments from east to west and concluded that they were derived from a land mass then lying east of the present peninsula.

Waldemar Lindgren made two short visits to the areas east of Ensenada in 1888 and 1889. Three articles resulted (Lindgren 1888, 1889, 1890), including the only petrographic descriptions published until those of Hirschi and de Quervain between 1927 and 1933 and Woodford and Harriss in 1938. Lindgren was the first to point out the many similarities between the geology of the Peninsular Ranges and that of the Sierra Nevada of California, U.S.A.

Emmons and Merrill (1894) published a comprehensive geologic report on the northern part of the peninsula. This paper synthesized the reports of Gabb and Lindgren and collated their work with that of Fairbanks (1893), Lawson (1893), and others in southern California, U.S.A.

In commenting on Lindgren's comparison of the Sierra Nevada and the Peninsular Ranges, Emmons and Merrill concluded (p. 512):

On the other hand, the general deduction that the depressed area of the Gulf of California and its continuation in the desert valleys to the north bears the same relation to the Peninsular Sierra that the Great Basin does to the Sierra Nevada appears to be borne out by what is known of the general structure of the former.

Wittich (1909) and Böse and Wittich (1913) made comprehensive inventories of the natural resources of the peninsula. With the beginning of the world-wide search for petroleum, Baja California was the subject of a series of reconnaissance surveys. Darton (1921) described the Tertiary and Upper Cretaceous strata of the peninsula but utilized names such as "Chico" and "Monterey" because of the similarity in age and rock type to previously recognized formations in California, U.S.A. Heim (1922) and Beal (Marland Oil Company of Mexico, 1924) were responsible for much of the stratigraphic nomenclature now used in the peninsula. Flores (1931) published the first geologic map of Baja California with a general geologic description.

Johnson (1924) reported the gulf islands to be mainly volcanic, and Hanna (1927) published an account of the "Geology of the West Mexican Islands."

Santillán and Barrera (1930) described the stratified rocks of the Pacific coast. Anderson and Hanna (1935) reported on the Upper Cretaceous rocks and fossils. Mina (1957) published a comprehensive study of the stratified rocks of the territory of Baja California with a map.

OUTLINE OF GEOLOGIC HISTORY

The rocks of Baja California contain the record of two geologic revolutions: we call the first the "mid-Mesozoic event" and the second the "mid-Cenozoic event." The mid-Mesozoic event occurred from Jurassic through middle Cretaceous time and is documented by volcanic strata of the island-arc type, regional metamorphism, and the emplacement of pervasive granitic rocks. It involved profound structural deformation, uplift, and erosion and probably resulted in fundamental geographic changes. The mid-Cenozoic event involved the accumulation of a wide variety of volcanic rocks, minor(?) metamorphism and granitic emplacement, and the creation of the Gulf of California, a mobile tectonic belt. This event, initiated at or before the beginning of the Miocene Epoch, is still going on.

These two events divide the history of Baja California into significant intervals. Everything that occurred before the culminating granitic emplacements of middle Cretaceous time is referred to as "prebatholithic"; everything thereafter, as "postbatholithic." Postbatholithic history is divided into events that occurred before the formation of the Gulf of California province (approximately 30 m.y. ago) and those that occurred afterward.

GEOLOGIC MAP OF BAJA CALIFORNIA

This memoir is accompanied by a geologic map of the state of Baja California (Pls. 1-A, 1-B, 1-C) published at a scale of 1:250,000. The principal source for the base map is the U.S. Army Map Service (1958-1964) series F 501, edition 1, Estados Unidos Mexicanos, 1:250,000. The degree of map precision (Fig. 3) varies from areas that have been mapped at scales as large as 1:10,000 to areas for which we have neither aerial photographs nor topographic maps so that the mapping there is by hilltop reconnaissance. All published and manuscript geologic maps known to us as of June 1970 (Fig. 2) were consulted and used in our compilation.

Several points should be made concerning the map legend. The distinction between Qf, Tpf, and Tmf and between Qb, Tpb, and Tmb is in many places based solely on the degree of dissection or deformation. Tmf is used throughout the peninsula for nonmarine strata underlying Miocene (in many places undated) volcanic rocks. In a few places, the unit may be younger than Miocene, and much of it could be as old as Oligocene, Eocene, Paleocene, or even Late Cretaceous. Tc is applied to ridge-capping conglomerate that lies on the otherwise-dissected old erosion surfaces cut on the basement rocks. Some of the conglomerate units are probably Eocene; others may be as old as Late Cretaceous or as young as Pliocene. For the prebatholithic volcanic terrane, Ka (Alisitos Formation) is used where fossil evidence indicates a Cretaceous age; Jv?, where fossil evidence indicates a Jurassic age; and *pb*, where the age can only be given as prebatholithic. No attempt has been made to draw boundaries between the areas of dated and undated rock. The designation *pbc* includes a varied assortment of calcareous, cherty, and, in some places, immature and volcanic strata that from scattered evidence apparently are of Carboniferous age.

The Instituto de Geología de México has a long-term objective of publishing a geologic map of the entire country at a scale of 1:100,000 on quadrangles bounded by 30' of latitude and 40' of longitude. Figure 3 gives the names of these quadrangle subdivisions for the state of Baja California. Each map sheet will be published with a text on the reverse side. The Tijuana-Presa Rodriguez and Bahía Santa Rosalia quadrangles have been submitted for publication, and several others may be submitted in the near future.

GEOGRAPHIC NAMES

The state of Baja California consists of about one-quarter of the vast tract of land known to the 17th century Spaniards as "California." By 1780, "California" had been divided into two political subdivisions: Alta (upper) California, with its

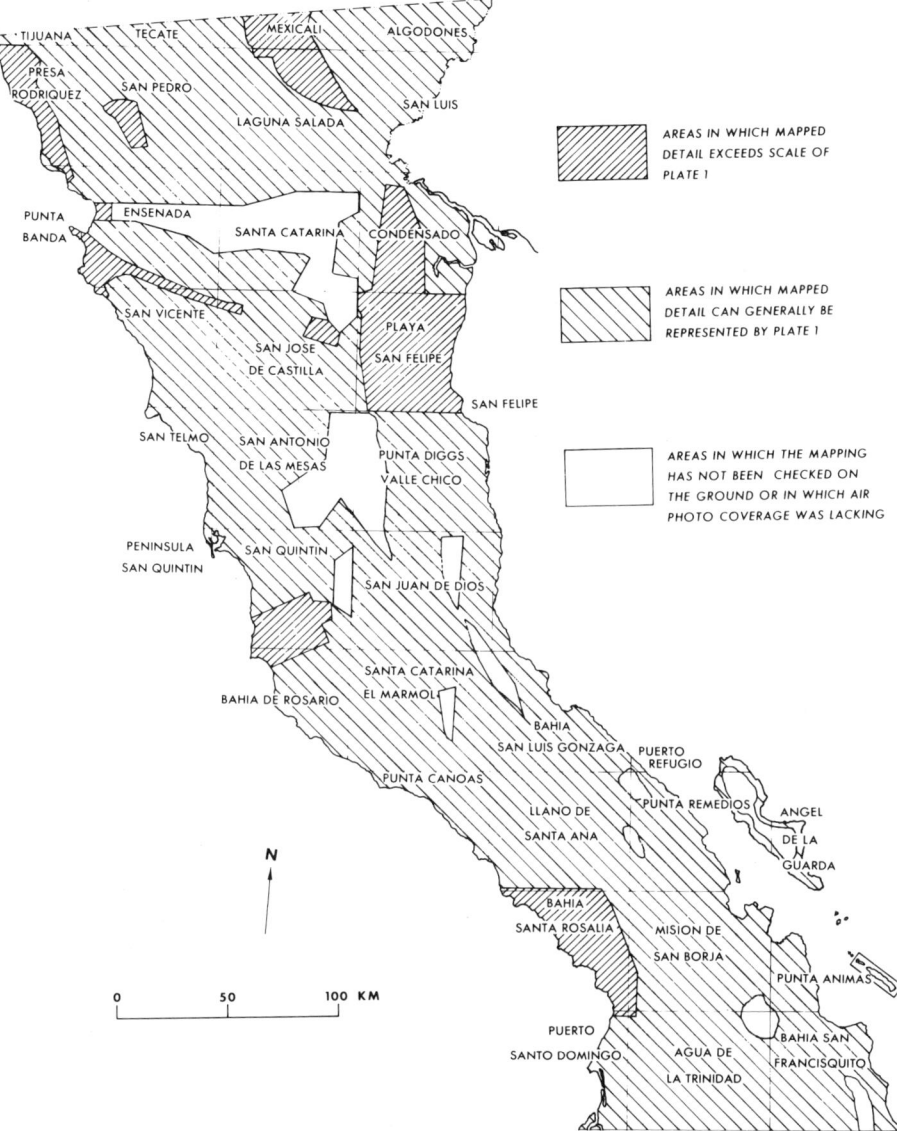

Figure 3. Index to geologic map reliability. Names of the 30′ × 40′ quadrangles are as designated by the Instituto de Geología de México.

capital at Monterey, and Baja (lower) California, with its capital at Loreto. When, in 1850, Alta California was admitted as the 31st state of the United States of America, it was called simply California. The adjective "Alta" is now just a historical memory. The portion of "California" belonging to Mexico has retained the name Baja California, but it too has been politically subdivided. On January 16, 1952 (Martínez, 1960, p. 552), the northern portion of Baja California was officially designated as the Estado (state) de Baja California and admitted as the 29th state in the Mexican Federation. The region south of lat 28° N. was designated the Territorio (territory) de Baja California.[1]

The political subdivisions of "California" do not correspond to the geologic subdivisions. It is often necessary to refer to conditions both north and south of the state of Baja California when discussing its geology. We have attempted to standardize our use of geographic names to minimize the confusion that can result from the similarities of the names of the three present political subdivisions of "California."

The state of California of the United States of America is referred to directly as California, U.S.A., or is implied by the mention of a county name, such as San Diego County, California. The designation "southern California" refers to that part of California, U.S.A., south of lat 35° N. This includes the Transverse Ranges, the Los Angeles basin, the Peninsular Ranges, the southern Mojave Desert, and the western Colorado desert.

The peninsula of Baja California, more simply, Baja California or just the peninsula, is used to designate the combined state of Baja California and the territory of Baja California, that is, all of Mexican California. We have often shortened "state of Baja California" to just "state" where the context is clear.

There is no agreement on the use of place names in Baja California. Existing maps differ considerably not only on spelling but on the names for features, and these map names are different from the names in local usage. For the most part, we have tried to use the names that have the greatest local recognition. The names of villages and ranches are usually the same as those found in Gerhard and Gulick's *Lower California Guidebook* (1970). Topographic features present a more confusing picture. The names on the available maps are often misleading or misplaced. North of lat 30° N., many of the names used are from the U.S. Army Map Service sheets (1958-1964), but in the southern part of the state, we have relied on local usage.

---

[1] On October 9, 1974, the territory of Baja California became the state of Baja California Sur.

2
Prebatholithic Terrane

The preintrusion rock patterns and geologic history of Baja California are poorly known because the prebatholithic rocks have been metamorphosed and intruded by an extensive belt of batholiths. Furthermore, many areas critical to the study are submerged beneath the Pacific Ocean and the Gulf of California.

We divide prebatholithic rocks of the state of Baja California into four belts, three on the peninsula (Fig. 4) and one to the west on the continental borderland. Localities cited in this chapter are shown in Figure 4.

The rocks of the westernmost belt are exposed only on some of the offshore islands, the Vizcaíno Peninsula to the south, and the Palos Verdes Peninsula of Los Angeles County. These rocks are similar to those of the Franciscan assemblage (Bailey and others, 1964) of northern and central California, U.S.A. Along the western shore of Baja California is a belt consisting of volcanic and volcaniclastic rocks. In the central part of the peninsula is a belt of metamorphosed shale and sandstone. On the eastern side is a belt that contains a great variety of metasedimentary rocks, including those derived from carbonate rock, chert, wacke, pebbly mudstone, arkose, quartz sandstone, and shale. These four belts are delineated only by rock type. We do not imply a sequential stratigraphic relation between rocks of these associations.

"FRANCISCAN" BELT

Rocks of the westernmost prebatholithic belt do not crop out in the state of Baja California but are presumed to underlie much of the continental borderland to the west. These rocks are commonly compared to the rocks of the Franciscan assemblage of central and northern California, U.S.A. (Bailey and others, 1964), and include graywacke, bedded chert, serpentinite, and glaucophane-, crossite-, and lawsonite-bearing schists. North of the international border, rocks of the Franciscan belt are exposed in the Channel Islands (Weaver, 1969) and on the Palos Verdes Peninsula (Bailey and others, 1964) and underlie the western part of the Los Angeles basin west of the Newport-Inglewood fault (Bailey and others, 1964).

South of lat 28° N. (the southern border of the state of Baja California), the

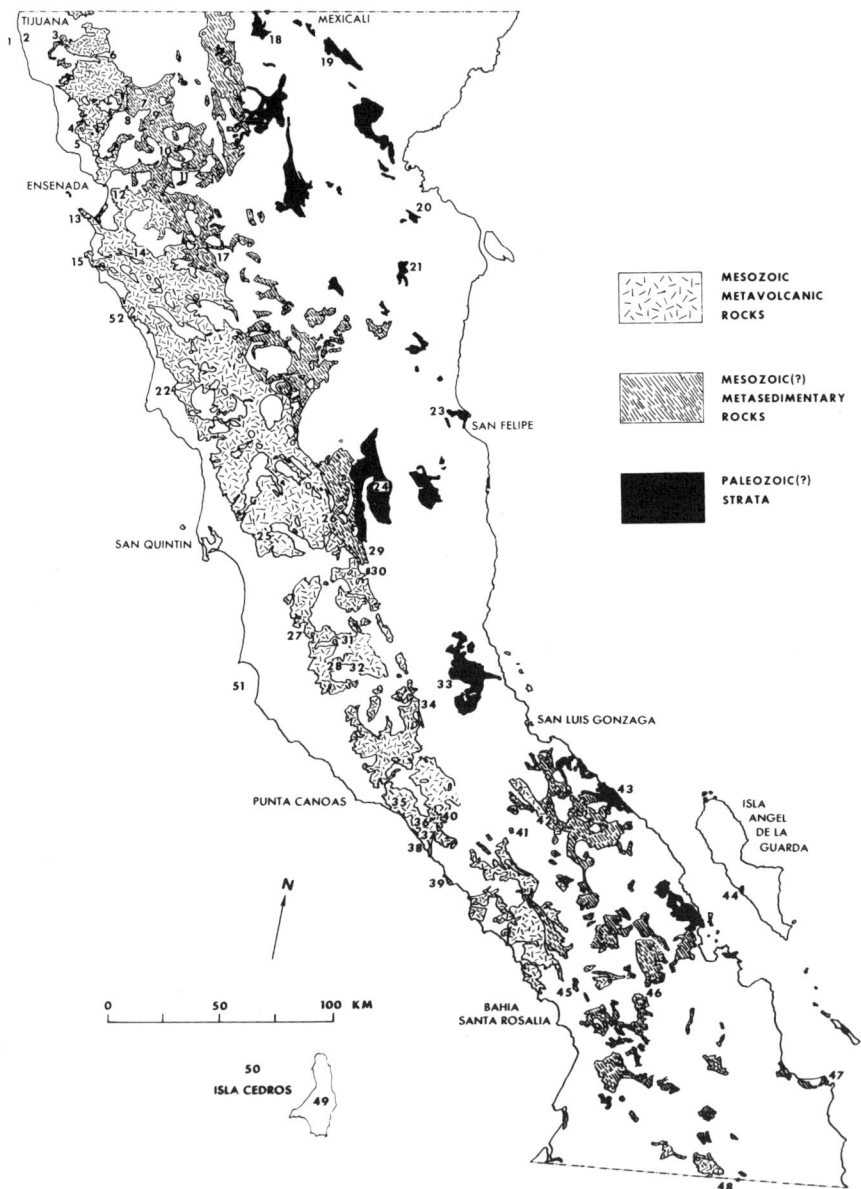

Figure 4. Distribution of prebatholithic rocks. Numbers 1 through 52 correspond to localities cited in the text.

association of ultramafic crystalline rocks, serpentinite, and sedimentary rocks like the Franciscan assemblage of California, U.S.A., crop out on the Vizcaíno Peninsula (Beal, 1948; Mina, 1957, p. 155) and the western border of Bahía Magdalena (Hirschi and de Quervain, 1933; Yeats and others, 1971).

West of the state of Baja California, rocks like the Franciscan are known from Isla Cedros (loc. 49; Hanna, 1925, p. 268) and the Islas de San Benito (loc. 50;

Cohen and others, 1963). Butler (App. 1) mapped exposures on the sea floor along the western edge of the continental escarpment west of Baja California and reported:

Discrete layers of red radiolarian chert fragments occur in the ooze and are lithologically similar to chert from the Franciscan [assemblage] of California. The chert fragments are believed to have been derived "in place" from Franciscan fault breccia.

During the Miocene Epoch, such rocks were extensively exposed in the offshore area. The existence of this offshore "Franciscan" highland was recognized by Lindgren (1890, p. 29). The detritus derived from the highland is best known for its occurrence in the Miocene San Onofre Breccia of southern California (Woodford, 1925). Such detritus is also present in the Miocene strata of the Islas los Coronados (loc. 1; Woodford, 1925; Emery and others, 1952) that lie just west of Tijuana. Minch (1967) reported finding "San Onofre-type clasts" in the Miocene Rosarito Beach Formation south of Playas de Tijuana (loc. 2).

Rocks that most closely resemble Franciscan strata in the coastal basement are found just north of Arroyo San José (loc. 37). Here, predominantly volcaniclastic Jurassic(?) strata, containing a deep-water fauna, include bedded chert derived from silicified volcanic ash (Minch, 1968, oral commun.). Cohen and others (1963) reported chert derived from silicified graywacke on the Islas de San Benito (loc. 50).

The age of the Franciscan-like rocks in the continental borderland is suggested by the following data. North of the international border, the Santa Cruz Island Schist is intruded by diorite with a K-Ar age of 145 m.y. (hornblende; Weaver, 1969). The relation of the Santa Cruz Island Schist to the glaucophane-facies schist found elsewhere in the Channel Islands is unknown. Suppe and Armstrong (1972) reported K-Ar ages for four rocks from the blue schist and chert terrane of Catalina Island: 95 ± 2.5 m.y. (blue amphibole), 107 ± 2 m.y. (white mica), 105 ± 2 m.y. (whole rock), and 109 ± 3 m.y. (hornblende). They determined an age of 108 ± 2 m.y. for blue schist from Sixty Mile Bank. From Isla de San Benito del Este, they reported two blue amphibole-bearing rocks with K-Ar ages of 148 ± 3 m.y. (white mica), 148 ± 5 m.y. (hornblende), and 104 ± 2 m.y. (blue amphibole). For rocks studied by Kilmer (1969) on Isla Cedros, Suppe and Armstrong (1972) reported a K-Ar age of 145 ± 6 m.y. (hornblende) for quartz diorite, and values of 94 ± 4 m.y. (blue amphibole), 109 ± 2 m.y. and 110 ± 2 m.y. (white mica) for rocks of the blue schist and chert Franciscan-like terrane. Yeats and others (1971) reported dates of 134 to 136 m.y. for related rocks from the Bahía Magdalena area, territory of Baja California. The quartz diorite is part of a volcanic-plutonic terrane believed to structurally overlie the Franciscan terrane.

In summary, Franciscan-like rocks, including glaucophane-facies schist and metamorphic rocks, are present in the continental borderland from the Channel Islands to Bahía Magdalena. The time of metamorphism of these rocks is not well established but was probably late Mesozoic.

VOLCANIC-VOLCANICLASTIC BELT

Mesozoic volcanic-volcaniclastic rocks crop out in the northern Santa Ana Mountains, southern California, U.S.A., where they have been called the Santiago Peak Volcanics (Larsen, 1948; Fife and others, 1967). Mesozoic volcanic-volcaniclastic rocks have been described from northernmost Baja California (Flynn, 1970), the area southeast of Ensenada (loc. 12; Schroeder, App. 1), Valle Santo Tomás (loc. 14; Santillán and Barrera, 1930; Allison, 1955; Silver and others, 1963), the area east of El Rosario (loc. 27; Reed, App. 1), Misión San Fernando (loc. 32; Beal, 1948; Böse and Wittich, 1913), Arroyo San José (loc. 37; Minch, 1969), the Bahía Santa Rosalía quadrangle (for location, see Fig. 3; Fife, 1969), and the region along the southern border of the state as far east as Miraflores (loc. 48). Throughout this distance, the rocks are almost continuously exposed. They nowhere cross the peninsula and only extend into the gulf drainage near Arroyo Calamajué (loc. 42).

No fossils have been found in the volcanic sequence between the Agua Blanca fault (loc. 14; south of Ensenada) and Los Penasquitos Creek, San Diego County. Extensive areas south of the fault have also proven unfossiliferous.

Petrified wood of undetermined age has been found in volcanic strata west of Hodges Reservoir, San Diego County (D'Vincent, 1967). These are probably continental strata.

Between the international border and La Misión (loc. 4), volcanic and pyroclastic dacite and andesite are predominant with basalt and rhyolite subordinate. Hawkins (1970b, p. 25) studied these rocks along the Ensenada-Tecate Highway (Mexican Highway 2) just south of the border and described them as follows:

The andesitic to dacitic rocks are almost exclusively lithic-crystal tuffs and must have been formed by rather explosive volcanic activity. They have been moderately to extensively altered and recrystallized and the original textures of the matrix are difficult to recognize. . . . Rocks derived from volcanic flows were not positively identified.

. . . All of the samples . . . show textures and mineralogic evidence for a post-consolidation phase of low P-T recrystallization. The characteristic mineral assemblages albite-epidote-chlorite-muscovite; albite-epidote-chlorite-biotite or albite-chlorite-actinolite (all with quartz) all are typical of metamorphism under greenschist facies conditions [App. 3].

Hawkins (1970b) also pointed out the chemical similarity between the volcanic rocks and the average composition of the Southern California batholith (Table 1) and suggested that "the vulcanism, plutonism and thermal metamorphism probably were essentially contemporaneous events and parts of a single magma generation cycle."

North of the Agua Blanca fault, sedimentary strata are uncommon. Those occurring east and southeast of La Misión (loc. 5) are predominantly volcaniclastic, varying from conglomerate to shale. The least metamorphosed strata are reddish or purplish brown, the slightly more metamorphosed strata are green. Southeast of Ensenada (loc. 12), Schroeder (App. 1) described a well-exposed, well-preserved section of pyroclastic rhyolite and dacite.

South of the Agua Blanca fault, limestone, calcareous siltstone, and mudstone are interbedded with volcanic sandstone, volcanic conglomerate, tuff, and volcanic breccia. The sedimentary rocks represent deep to shallow and marine to nonmarine depositional environments. The sequence ranges from coarse sedimentary breccia to clean limestone and from basalt to rhyolite. Andesite is the predominant volcanic rock. The entire sequence is many thousands of meters thick. It varies widely in stratigraphic components from area to area. No two measured stratigraphic sections are alike, and additional mapping and stratigraphic work are required before subunits can be recognized and correlated.

Allison (1955) redescribed the type section (Fig. 5) of the Alisitos Formation (Santillán and Barrera, 1930) at Rancho Alisitos in Valle Santo Tomás (between locs. 14 and 15).

Students and faculty from California Institute of Technology have mapped and studied the Alisitos Formation near Punta Cabra (loc. 52; L. T. Silver, 1968, written commun.). Silver and others (1969) reported an isotopic uranium-lead date of 127 ± 5 m.y. for zircon from a volcanic rock that is contemporaneous with the Alisitos Formation.

Reed (App. 1) described a very different section of Alisitos age (Fig. 6) east of El Rosario (loc. 27). The Albian-Aptian strata in the area north of Misión San Fernando (loc. 31) contain prominent units of massive gray limestone interbedded with several varieties of calcareous sandstone. Unfortunately, the fossil preservation in this area is poor because of the proximity of granitic intrusions. During the reconnaissance geologic mapping carried out by the Consejo de Recursos Naturales no Renovables (1965), a number of paleontologic collections were made. From these, Perrilliat-Montoya (1968) described collections obtained from the excellently preserved and previously unreported section south of Mina Ursula (loc. 28), from the extreme eastern margin of Mesozoic volcanic rocks 6 km north of Rancho El Rosarito (loc. 26), from 10 km north of Misión San Fernando (loc. 31), and from Arroyo del Pabellón 15 km east of San Quintín (loc. 25).

In the area south of Punta Canoas (loc. 35), the middle Cretaceous strata are

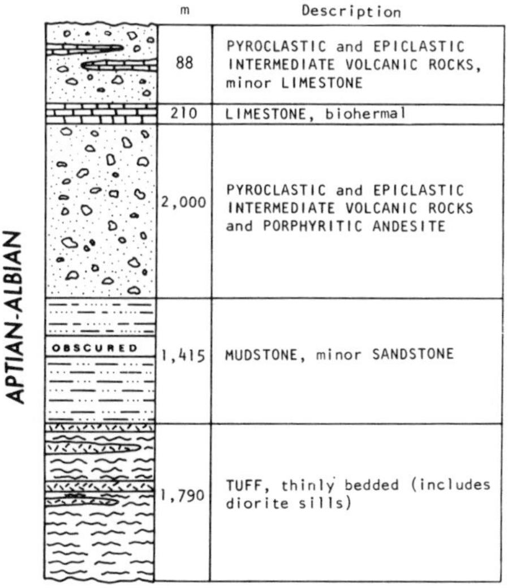

TABLE 1. CHEMICAL ANALYSES OF PREBATHOLITHIC VOLCANIC ROCKS

Sample	Baja 14a	Baja 12a	Baja 13c	SCB	*
SiO_2	51.66	70.29	71.35	64.1	67.4
TiO_2	1.86	0.42	0.31	0.6	0.6
Al_2O_3	15.12	14.30	13.58	16.5	13.7
Fe_2O_3	2.73	0.82	1.08	1.3	1.4
FeO	9.91	3.24	2.77	3.8	4.2
MnO	0.25	0.10	0.11	0.04	0.1
MgO	3.68	0.61	0.64	2.4	1.2
CaO	8.12	3.32	2.11	5.2	3.5
Na_2O	4.15	3.64	4.33	3.3	4.3
K_2O	0.35	1.18	2.03	2.0	1.7
H_2O^+	1.85	1.45	0.91	0.5	..
H_2O^-	0.35	0.26	0.16	..	..
P_2O_5	0.20	0.12	0.11	..	..
CO_2	n.d.	n.d.	0.38	..	..
Total	100.23	99.75	99.87	99.7	98.1

Baja 12a — Dacitic lithic-crystal tuff (recrystallized in albite-epidote hornfels facies) Tijuana - Tecate road at 1,266 km marker

Baja 13c — Dacitic crystal tuff (recrystallized in albite-epidote hornfels facies) Tijuana - Tecate road at 1,264 km marker

Baja 14a — Diabasic basalt (flow?) (recrystallized in albite-epidote hornfels facies) Tijuana - Tecate road at 1,262 km marker

SCB — Average composition of Southern California batholith (Larsen, 1948)

* Rough approximation of an "average" composition for Tijuana-Tecate area volcanic series based on composition of Baja 14a and Baja 13c using proportions of 1:4.

NOTE: modified from Hawkins, 1970b.

Figure 5. Generalized stratigraphic section (after Allison, 1955) of the Alisitos Formation in the type area near Rancho Alisitos, Cañon Santo Tomás (between locs. 14 and 15, Fig. 4). Overlain by Rosario Formation. Thickness in meters. Base not exposed.

Stratigraphic column (APTIAN-ALBIAN):
- 88 m: PYROCLASTIC and EPICLASTIC INTERMEDIATE VOLCANIC ROCKS, minor LIMESTONE
- 210 m: LIMESTONE, biohermal
- 2,000 m: PYROCLASTIC and EPICLASTIC INTERMEDIATE VOLCANIC ROCKS and PORPHYRITIC ANDESITE
- 1,415 m: MUDSTONE, minor SANDSTONE (OBSCURED)
- 1,790 m: TUFF, thinly bedded (includes diorite sills)

principally fine-grained siliceous siltstone or tuff(?) containing *Acila* sp., *Nanonavis* sp., and unidentified ammonoids and echinoids. The fauna and rocks resemble the Alisitos strata west of San Telmo (loc. 22). Along Arroyo Cuervito (loc. 36) just north of Arroyo San José, these middle Cretaceous strata rest unconformably on a section of fossiliferous pre-Alisitos (possibly Jurassic) strata at least 1,600 m thick (Minch, 1969).

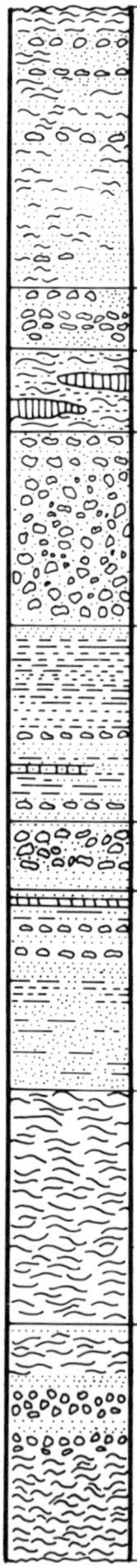

m	Description
325	TUFF, AGGLOMERATE, VOLCANIC BRECCIA, and ANDESITE FLOW: Tuff is green to purple, thin-bedded to massive, and moderately to well-indurated. Agglomerate is dark brown to purple, massive, well-indurated, with basalt and andesite clasts. Breccia is brown to purple, massive, well-indurated. Andesite flows are purple, commonly vesicular, and porphyritic. Overlain by Rosario Formation
50	AGGLOMERATE and TUFF: Tuffaceous agglomerate is purple, massive, moderately indurated, with clasts of tuff. Lapilli tuff is green to purple, thin-bedded, and poorly to moderately indurated
70	DACITE, ANDESITE, and BASALT FLOWS, and AGGLOMERATE: Volcanic flows are light to dark gray-brown, and purple; many are porphyritic and vesicular. Agglomerate is purple to green, massive, moderately indurated, with clasts of tuff and andesite breccia
210	VOLCANIC BRECCIA, AGGLOMERATE, and TUFF: Entire sequence is green, massive to thin-bedded, and poorly to well-indurated. Clasts of intermediate composition volcanic rock up to one meter in diameter
165	VOLCANIC SHALE, SILTSTONE, SANDSTONE, and locally CONGLOMERATE, and LIMESTONE: Shales are medium brown to black, well-indurated, shaley to thin-bedded, carbonaceous, and fossiliferous (ammonites). Siltstones are brown, thinly bedded, moderately indurated, and commonly calcareous. Sandstones (graywacke, feldspathic and arkosic arenite) are tan to medium brown, fine- to coarse-grained, poorly to fairly well sorted, massive to thinly cross-laminated, locally calcareous. Limestone is gray, volcaniclastic, fossiliferous (*Ostrea*)
60	VOLCANIC WACKE and LITHIC ARENITE and CONGLOMERATE: Sandstones are greenish-gray and light brown, massive to thin-bedded, commonly poorly sorted, well-indurated, fossiliferous (*Lima*, *Ostrea*). Cobble to boulder conglomerate is greenish gray, massive, well-indurated, with clasts of dacite and andesite
50–500	VOLCANIC WACKE and SUBARKOSE ARENITE, SILTSTONE, CONGLOMERATE, and LIMESTONE: Sandstone is green, brown, or gray, medium- to coarse-grained, massive to thin-bedded, well-indurated, and fossiliferous (*Quadratotrigonia*, *Pterotrigonia*, *Apiotrigonia*). Tuffaceous siltstone is greenish-brown, thin- to thick-bedded, poorly to moderately indurated, and fossiliferous (echinoids). Conglomerate is brown, massive, and well-indurated. Limestone is light gray, volcaniclastic, fossiliferous (*Nerinea*)
275	DACITE FLOWS: light to medium gray and porphyritic
	Undetermined thickness of DACITE FLOWS, AGGLOMERATE, VOLCANIC BRECCIA, and WELDED TUFF: Flow units are light gray, and porphyritic. Agglomerate and breccia units are gray to green, massive, and well-indurated. Tuff units are pink, well-indurated, and massive

ALBIAN–APTIAN

Figure 6. Stratigraphic section of the Alisitos Formation east of El Rosario (loc. 27, Fig. 4; from Reed, App. 1). Thickness in meters. Base not exposed.

Albian-Aptian fossils have been collected from several sections within the Bahía Santa Rosalía quadrangle (Fife, App. 1), where the fossiliferous rocks rest on thousands of meters of strata that have so far proven unfossiliferous (Fig. 7). No fossils have been found in the prebatholithic rocks south of the Agua Refugio area (lat 28°47' N., loc. 45) except some bivalve casts in the prebatholithic volcaniclastic strata west of locality 48 (D. Barthelmy, 1974, oral commun.).

Lindgren (1888, 1889) was first to clearly recognize the Mesozoic volcanic rocks. He observed them as a discontinuous belt extending some 30 km inland from the international border to as far south as Punta Banda (loc. 13), the southern extent of his visit. He correctly reported that the volcanic rocks underlie the unmetamorphosed Cretaceous strata. The low grade of metamorphism in many of the Mesozoic volcanic rocks led Lindgren to believe that these overlie the granitic rocks.

In 1892 Merrill (1897) traversed the peninsula through El Rosario and Santa

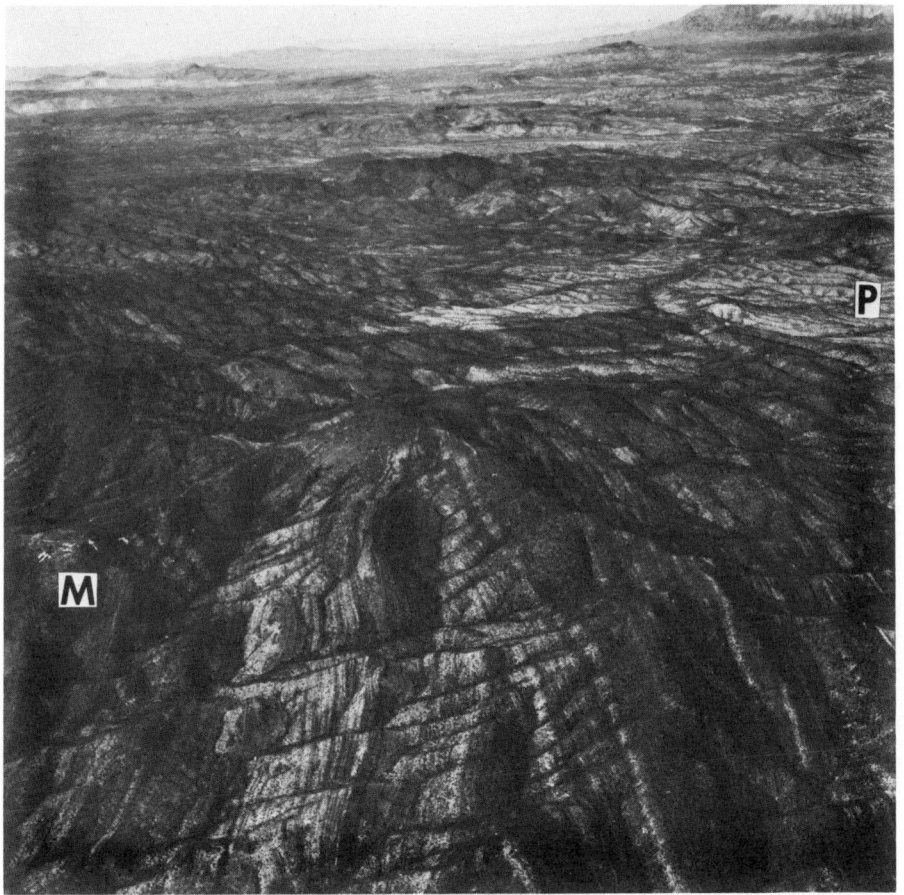

Figure 7. Anticline in prebatholithic strata, Santa Rosalía quadrangle. Oblique aerial view north from the vicinity of Mina Columbia (on ridge above M) across the Llano de Santa Ana (P). In the foreground is unfossiliferous, weakly metamorphosed siltstone and sandstone of pre-Alisitos age. Paleocene nonmarine strata mantle the irregular basement-rock surface of the Llano de Santa Ana. Photograph by John Shelton.

Catarina. He too recognized the "pre-Chico eruptives" including the calcareous and fossiliferous section near Misión San Fernando (loc. 32), later used as a type area by Beal (Marland Oil Company of Mexico, 1924), and the section of overlying rhyolite more recently described by Reed (App. 1; Fig. 6). Merrill considered these rocks to be of Mesozoic or Paleozoic age.

Böse and Wittich (1913) and Wittich (1909) were the first to recognize, on paleontologic evidence, that the volcanic and associated sedimentary strata underlying the unmetamorphosed Upper Cretaceous rocks were at least in part also of Cretaceous age and were intruded by granitic rocks. Darton (1921, p. 725-726) wrote as follows:

A large part of the peninsula is underlain by Cretaceous rocks which outcrop extensively in the central and northern parts. Two principal series are present, both of late Cretaceous age and separated by an unconformity, the older series, of unknown correlation, having been uplifted, flexed, and cut by large igneous masses before the younger series, Chico, was deposited.

These pre-Chico Cretaceous rocks consist of conglomerates, quartzites, tuffs, and agglomerates with large bodies of interbedded eruptive rocks. They are also cut by dikes and large stocks of igneous rocks of various kinds. In many localities the igneous rocks predominate over the sediments or pyroclastics, and in places there is much metamorphism. Unaltered or but little altered sandstone and shales appear in places. . . . Limestone also occurs. It is conspicuous north and northeast of the ruins of Mission San Fernando, 30 miles due east of Rosario. . . . The limestone is filled with fossil oysters of upper Cretaceous age.

Despite his recognition of Cretaceous intrusions and metamorphosed Cretaceous strata, Darton (1921) apparently believed (as did Merrill, 1897) that the major plutons of the peninsula were pre-Cretaceous in age.

Beal (1948) was the first to apply a name, San Fernando Formation, to the Cretaceous volcanic strata. In 1948 he wrote (p. 38-39):

The irregularly altered rocks, for which the name San Fernando formation was proposed by the writer (Anon. [Marland Oil Company of Mexico], 1924) from the prominent exposures north of San Fernando, include slate, conglomerate, quartzite, limestone, and volcanic rocks, with some sandstone and shale. Darton (1921), whose work was carried on concurrently with that of the present expedition, referred to these rocks as "pre-Chico Cretaceous rocks."

Little is known of the stratigraphic limits of the San Fernando formation; its actual contact with the granitic batholith has not been studied, and its relationship to the older metamorphic rocks is unknown.

Santillán and Barrera (1930) gave the name Alisitos Formation to the strata (Fig. 5) at Rancho Alisitos in Valle Santo Tomás (between locs. 14 and 15). They dated the sequence as follows:

Todos los fósiles que colectamos en este exploración fueron bondadosamente clasificados por el Dr. B. L. Clark, professor de la Universidad de California, E.U.A. Estos fósiles fueron clasificados como de la formación Paskenta, que corresponde . . . a la parte inferior del Cretácico medio de nuestro país.

Woodford and Harriss (1938), describing the rocks west of the Sierra San Pedro Mártir, called the prebatholithic Cretaceous strata the San Telmo Formation. In this area, Cretaceous fossils were found within the volcanic strata almost to the granitic contact, and there is no question that the batholiths are relatively younger.

Allison (1955) described the Punta China area (loc. 15), 6 km west of Rancho Alisitos, and adopted Santillán and Barrera's name of Alisitos Formation. At that time, he suggested a middle Albian age for at least the upper member and a similar or possibly older age for the lower member. In 1964, however, he indicated an Aptian age (p. 13) as follows:

Early Cretaceous.—Youngest rocks intruded by the Peninsular Range batholith are Aptian and possibly younger. No post-Jurassic and pre-Aptian rocks are recognized, although possible areas of outcrop can scarcely be considered as geologically explored. An abundantly fossiliferous section of Lower Cretaceous prebatholithic rocks, the Alisitos Formation, occurs in the northwestern part of Baja California (Santillán and Barrera, 1930; Allison, 1955; Allen and others, 1961). It consists of more than 5,000 meters, possibly twice that thickness, of predominantly andesitic pyroclastic rocks and immediately derived sedimentary rocks. The lower part of the section is commonly siliceous, finely clastic, and thinly bedded, with Aptian ammonoids and other invertebrates representative of sand- to mud-bottom communities. Higher parts of the section are generally more coarsely pyroclastic and epiclastic, apparently derived from a near volcanic source. Fossils from the upper part of the section are much more abundant and diverse, and include a rich representation of Aptian pantropical hermatypic coral and pachydontid bivalve reef ("Urgonian") faunas. Species known from California and more northern areas of outcropping Aptian and Albian rocks are scarce in Alisitos faunas. Uppermost strata of the Alisitos Formation in its type area, with a rich Ostracoda fauna of uncertain temporal significance, are of nonmarine origin. A specimen of the ammonoid *Douvilleiceras*, identified by J. B. Reeside, Jr. as close to *D. mammillatum* (Schlotheim), has been collected in the same general area of northwestern Baja California (Larsen and others, 1958, p. 46) and is indicative of Alisitos rocks slightly younger (Albian) than those of the type area of the formation. . . .

Rocks similar to those just described can be traced southward to almost the center of the Baja California Peninsula. Briefly studied exposures have been recognized for many years near the site of the old Dominican Mission of San Fernando, about 325 kilometers south of the northern boundary of Baja California. This is the type locality for the San Fernando Formation (Anon. [Marland Oil Company of Mexico], 1924; Beal, 1948). The distinct lithological and paleontological similarities of the San Fernando and Alisitos Formations, in their type areas, with the almost continuous outcrop of comparable rocks in intervening areas, require that the earlier but repeatedly preoccupied name, "San Fernando," be dropped in favor of Santillán and Barrera's incontrovertible name, Alisitos.

Silver and others (1963) demonstrated the continuity of "Albian" strata from the Alisitos type area to the San Telmo type area and recognized the priority of the name Alisitos. They also recognized the probable correlation with the San Fernando Formation in its type locality but, lacking demonstrable continuity, left this correlation for future work.

Minch (1969) found that fossiliferous strata in Arroyo San José (loc. 37) are unconformably overlain by the Albian-Aptian Alisitos Formation. These pre-Alisitos (possibly Jurassic) strata (Fig. 8) consist of at least 1,600 m of pyroclastic and epiclastic volcanic rocks with a varied marine fauna. Similar strata are found at Punta Blanca (loc. 39).

The only definitely identified fossils in the pre-Alisitos strata are *Otapiria* sp. (probably *O. tailleuri*) of Late Triassic through Middle Jurassic age and Myophorallinae (probably genus *Linotrigonia*) of Late Jurassic to Cretaceous age. In the same beds are poorly preserved ammonoids that have been identified as middle

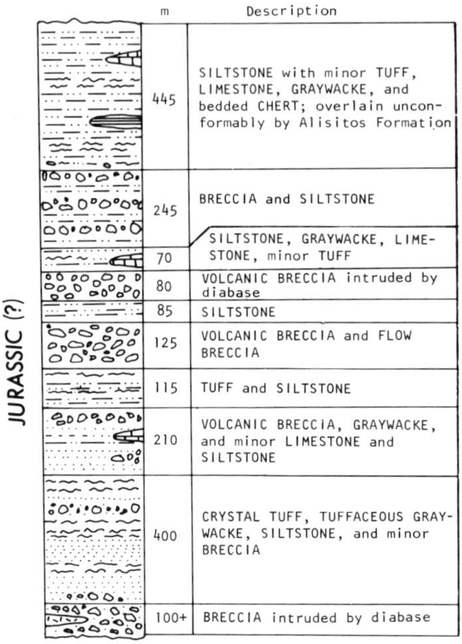

Figure 8. Section of pre-Alisitos strata exposed in Arroyo San José (loc. 37, Fig. 4). The lower part of the section is intruded by diabase. Thickness in meters. Base not exposed.

JURASSIC (?)	m	Description
	445	SILTSTONE with minor TUFF, LIMESTONE, GRAYWACKE, and bedded CHERT; overlain unconformably by Alisitos Formation
	245	BRECCIA and SILTSTONE
	70	SILTSTONE, GRAYWACKE, LIMESTONE, minor TUFF
	80	VOLCANIC BRECCIA intruded by diabase
	85	SILTSTONE
	125	VOLCANIC BRECCIA and FLOW BRECCIA
	115	TUFF and SILTSTONE
	210	VOLCANIC BRECCIA, GRAYWACKE, and minor LIMESTONE and SILTSTONE
	400	CRYSTAL TUFF, TUFFACEOUS GRAYWACKE, SILTSTONE, and minor BRECCIA
	100+	BRECCIA intruded by diabase

Cretaceous (M. Murphy, 1968, oral commun.) and belemnoids tentatively identified as Late Jurassic or Early Cretaceous (J. A. Jeletsky, 1969, written commun.). These strata we refer to as Jurassic(?).

To the north, in San Diego County, M. A. Hanna (1926) assigned the name Black Mountain Volcanics to prebatholithic volcanic and volcaniclastic strata in the La Jolla quadrangle. However, the name Black Mountain was preoccupied, and Larsen (1948) proposed the name Santiago Peak Volcanics and designated Santiago Peak in the Santa Ana Mountains, Orange County, as the type locality.

Both Hanna and Larsen assigned a Jurassic age to these strata because they overlie the Bedford Canyon Formation (then believed to be of Triassic age) and are intruded by the batholithic rocks (then assumed to be of Late Jurassic to Early Cretaceous age). Fife and others (1967) reported the discovery of Upper Jurassic (Portlandian) fossils within these rocks.

We use Santiago Peak Formation (Fife and others, 1967) for demonstrably Jurassic (Portlandian) strata in the northern part of the belt, to date only found north of the border; Alisitos Formation for all demonstrably Albian and Aptian strata; and Jurassic(?) for the pre-Alisitos strata of the San José area. On the 1:250,000 map (Pls. 1-A, 1-B, 1-C), we use the symbol *pbv* (prebatholithic, undifferentiated volcanic rocks) to designate the extensive areas in which the age is unknown.

Volcanic Dikes

Large swarms of Mesozoic volcanic and hypabyssal dikes are found just east of the Mesozoic volcanic belt. One swarm extends from south of Valle de las Palmas (loc. 6) in a continuous belt east of Valle Guadalupe (loc. 8) through Valle San Rafael (loc. 11) to southeast of El Alamo (loc. 17; Snyder, 1970). Rocks in this swarm range in composition from basalt to rhyolite. A second swarm, which extends from south of San Augustín (loc. 34) to the vicinity of Laguna

Chapala (loc. 41), consists of andesitic dikes. Both dike swarms cut most of the granitic rocks but are themselves metamorphosed. Between Valle de las Palmas and Valle Guadalupe, all dikes are clearly truncated by a large granodiorite-tonalite pluton. Some of the dacite dikes contain spherulitic and shard structures and show sedimentary structures analogous to those found in sandstone dikes. It is likely that the dikes were feeders to eruptions intermediate in age between the older and younger granitic intrusions. Similar dikes were reported by Lindgren (1888, p. 188), Woodford and Harriss (1938), and Larsen (1948).

SHALE-SANDSTONE BELT

East of the volcanic belt, through the Peninsular Ranges province of southern California and northern Baja California, we find a belt of variously metamorphosed quartz-bearing sandstone, argillite, and minor carbonate rocks.

Rocks belonging to this belt are best known north of the international border. Smith (1898) and Willis and Stose (1912) reported what were then believed to be Triassic fossils in slate and quartzite of the Santa Ana Mountains. Smith used the name Santa Ana Limestone for the limestone within these rocks, and Merrill (1914, p. 9, 10) used the term "Santa Ana metamorphic strata." As the name Santa Ana was preoccupied, Larsen (1948, p. 19) adopted the name Bedford Canyon Formation. He believed that many of the more severely metamorphosed rocks found elsewhere in the northern Peninsular Ranges were correlative with this formation.

Silberling and others (1961) reported the discovery of a Callovian fossil locality in Ladd Canyon adjacent to the type area of the Bedford Canyon Formation. They also reported a determination made by D. V. Ager that suggested that the fauna from the type locality in Bedford Canyon might also be of Late Jurassic age. Imlay (1963) reported Callovian pelecypods and ammonoids and (1964) the Bajocian ammonoid *Dorsetensia* from the Santa Ana Mountains localities once believed to be of Triassic age. Moscoso (App. 1) mapped the area, which included all of the above fossil localities. He was able to distinguish at least three major stratigraphic units including the carbonate and chert-bearing Oxfordian strata in Bedford Canyon, the carbonate-bearing Callovian strata in Ladd Canyon, and a thick sequence of flysch strata lying between these two units. Sedimentary structures indicate that the entire section is upside down. Farther east in the Winchester Quarry area, Webb (1939) reported a Mississippian coral. The fossil, however, was found in float, and Schwarcz (1969) concluded that it was not of local origin. Schwarcz correlated the rocks of the Winchester area with the Bedford Canyon Formation of Larsen. In 1968, M. A. Murphy (1969, written commun. to Allison) discovered fossil bivalves near Sun City immediately adjacent to the area mapped by Schwarcz. The fossils (Lamb, 1970) proved to be of Triassic(?) age.

Metamorphosed shale and sandstone of eastern San Diego County have in recent years been mapped, for want of more definitive information, as Julian Schist, with the town of Julian as the type locality. The unit was first designated as the Julian group (Merrill, 1914), then as the Julian series (Hudson, 1922), and finally as Julian Schist (Donnelly, 1935). In addition to metamorphosed sandstone, shale, and conglomerate, the unit includes small bodies of amphibolite and marble. An impression of an ammonoid found in the Julian Schist was identified by J. P. Smith as a Triassic form (Hudson, 1922). Unfortunately, the whereabouts of this fossil is unknown so that it is now impossible to re-examine the Triassic identification. No other fossil has ever been reported from the Julian Schist.

The schist, quartzite, and minor carbonate rocks of the western Sierra Juárez are very similar to the Julian Schist. Thin alternating layers of sand and shale are typical. Thousands of meters of metamorphosed argillaceous sandstone and shale crop out in the areas north and east of Valle Guadalupe. Here, however, graywacke is also an important constituent, and thin layers of carbonate rock are numerous.

Bell (App. 2) demonstrated that the metamorphosed slate-sandstone sequence north of Valle Guadalupe (loc. 7) lies unconformably beneath the volcanic-volcaniclastic sequence. This critical contact relation has not been precisely located elsewhere south of the Santa Ana Mountains. Figure 9 illustrates an unusual section of rocks mapped by Kaiser (App. 2) southeast of El Burro (loc. 9). This slate-quartzite-carbonate association is unknown elsewhere in the western part of the peninsula and may be an isolated exposure of Paleozoic strata.

The only author to distinguish specifically the rocks of the shale-sandstone belt in Baja California was Lindgren (1889, p. 14), who referred to it as the "slate series" and described it at Real del Castillo (loc. 10) as consisting of chlorite slate (metamorphosed mafic igneous rocks), quartzite, and carbonaceous slate. The chlorite-bearing rocks that Lindgren found in that particular area are atypical of the belt.

Between El Alamo (loc. 17) and the southern Sierra San Pedro Mártir (loc. 30) are extensive exposures of coarse schist and granitoid gneiss. These are probably metasedimentary rocks, but some of them may be metamorphosed igneous rocks. Similar terranes are found in San Diego County (Everhart, 1951; Merriam, 1959), where they are interpreted as being the "migmatization" of Julian Schist or the intimate mixing of metasedimentary and igneous components. In the Sierra San Pedro Mártir, Woodford and Harriss (1938, p. 1310) called this type of rock the Santa Eulalia Formation:

On the San Pedro Mártir plateau and along its western margin the plutonic masses are enveloped by broad belts of gneisses and other coarsely crystalline metamorphic rocks. A rather small part of these are highly siliceous quartz-biotite gneisses, and a still smaller portion feldspathic quartzite and crystalline limestone with lime silicates. Still other rocks are fine-grained schists. But the main types are coarsely banded quartz-plagioclase-biotite gneiss and similar rocks, many with garnet. Some of these gneisses are more than half quartz; others are close to quartz diorite and granodiorite in composition. Pegmatites are numerous, and some of the gneisses are obviously of injection origin. The type area is from Santa Eulalia north to the vicinity of Ciénega Santa Rosa.

Woodford described gneiss that contains as much as 75 percent quartz and other metamorphic rocks with inch-long sheaves of sillimanite. It seems plausible that the Santa Eulalia Formation is the higher grade equivalent of Lindgren's slate series.

A quartzite-phyllite sequence at the south end of the Sierra San Pedro Mártir closely resembles the Julian Schist of San Diego County. The road east of Rancho Nuevo crosses such a section (loc. 29). Other good exposures are the slate of the Sierra la Asamblea (east of loc. 42), the quartzite of the Laguna Chapala area (loc. 41), and the slate and quartzite exposures along the road between Rancho Todos Santos and Rancho San José (loc. 40). Similar rocks are widely exposed north and south of San Borja (loc. 46) and discontinuously exposed southeastward to Punta San Francisquito (loc. 47).

In all these areas, the only rocks that appear sufficiently unmetamorphosed to allow fossil preservation are northeast of Valle Guadalupe (locs. 7 and 9). Without fossils, it will never be easy to establish the stratigraphic identity of the shale-sandstone belt. Analogy to areas north of the international border suggests that the rocks are of Triassic and Jurassic age.

The rocks of the shale-sandstone belt were derived from a quartz-rich terrane that contained zircons of Precambrian age (Bushee and others, 1963). In contrast, sandstone of the Jurassic and Cretaceous volcanic belt is almost devoid of quartz, and it must be concluded that the sediments were derived from a wholly volcanic terrane.

PALEOZOIC METASEDIMENTARY BELT

The presence of carbonate-rich sequences, presumably of Carboniferous age, in the desert ranges north of the international border is well known. They occur in the Santa Rosa Mountains (Wright, 1946), around Borrego Valley (Sharp, 1967), and in the Coyote Mountains (Dibblee, 1954).

Hirschi (1926, p. 35) apparently was the first to recognize that the metamorphic rocks on the gulf side of the peninsula were unlike those of the Pacific side. The prebatholithic terranes both west and east of Laguna Salada (locs. 18 and 19) are distinctive for their relative abundance of carbonate rock (Barnard, 1968b). The basement rocks of the northern Sierra Pinta (loc. 20) are composed of metamorphosed shale, sandstone, conglomerate, and limestone, with some volcanic rocks and some banded chert (McEldowney, App. 1; Figs. 10 and 11). The southern Sierra Pinta (loc. 21) contains weakly metamorphosed pebbly mudstone, wacke, quartzite, thick carbonate units, banded chert, marlstone, and coarse arkose. The section at San Felipe (loc. 23) consists of alternate beds of marble and metasandstone with minor metashale and chert. At the eastern foot of the southern Sierra San Pedro Mártir (loc. 24), the section is predominantly metacarbonate rock, now recrystallized to wollastonite and other calcium silicate species. East of El Mármol

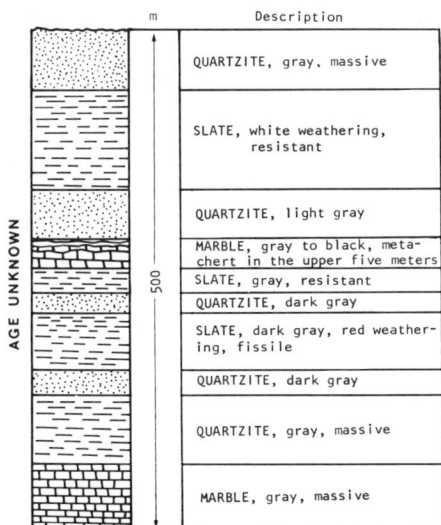

Figure 9. Stratigraphic section southeast of El Burro (loc. 9, Fig. 4; based on information from Kaiser, see App. 2). Top of section is terminated by a fault. Section rests on a thick sequence of slate and graywacke.

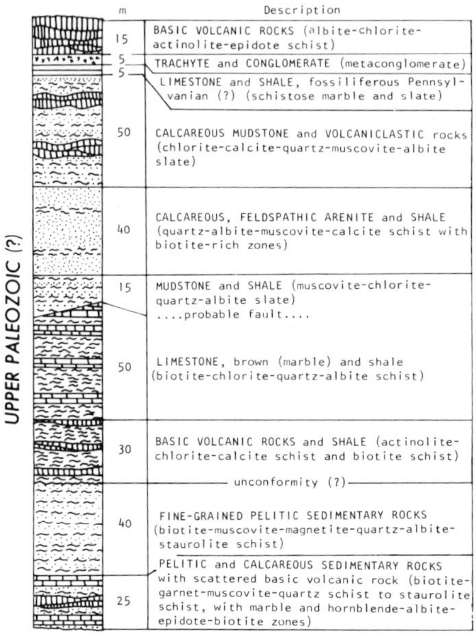

Figure 10. Generalized stratigraphic section of the metamorphic rocks in the northern Sierra Pinta (loc. 20, Fig. 4; after McEldowney, see App. 1). Rock types in parentheses are metamorphic equivalents as seen in this section. The section is unconformably overlain by Tertiary volcanic strata. Thickness in meters. Base not exposed.

(loc. 33; Fig. 12) is a large area of exposed metamorphic rocks that almost reaches the gulf coast at a travertine quarry north of Bahía San Luis Gonzaga. This terrane includes marble, slate, wacke, and pebbly mudstone. Along the coast south of Arroyo Calamajué are carbonate rocks, quartzite, and bedded cherts. Still farther south (loc. 43), this assemblage of rocks includes anhydrite. The anhydrite-carbonate-chert association is also found on Isla Angel de la Guarda (loc. 44) and Isla Datil just south of Isla Tiburón. Bedded chert is also found on the Sonora coast just north of Punta Chueca (opposite Isla Tiburón).

In 1969, McEldowney (1970) made the first reproducible discovery of Paleozoic fossils in the Peninsular Ranges. Crinoids, cup corals, and bivalves were discovered in a thin layer in the northern Sierra Pinta (loc. 20). The age was determined as "probably Carboniferous" (R. L. Langenheim and G. D. Webster, 1970, oral commun.). Crinoid columnals in a pebble layer and nautiloids and other forms in a limestone bed were discovered later in two localities northeast of El Mármol (loc. 33; Fig. 12).

A basal conglomerate bed in the Sierra Santa Rosa, underlying the Miocene volcanic rocks and overlying granitic basement, includes a variety of coarse, angular, exotic clasts. Prominent among these are boulders of fossiliferous limestone (Gastil and others, 1973). These blocks, as much as 30 cm across, cannot have been transported a great distance. David LeMone (1971, written commun.) determined that the age of the fossils is Early Permian.

PROBLEMS OF NOMENCLATURE

Some of the formation names mentioned in this chapter have recently been used very differently. Thus, the Consejo de Recursos Naturales no Renovables

(1965) uses the name Formación San Fernando (following Beal) for the rocks Woodford called San Telmo and we call Alisitos, Formación San Telmo for those rocks that Lindgren called "slate series" and Woodford called Santa Eulalia Formation, and Formación Santa Eulalia for those portions of the batholithic terranes that the Consejo interprets to be of metamorphic origin. This usage has not been followed here because the name San Fernando is preoccupied (Allison, 1964) and because the names San Telmo and Santa Eulalia were used differently by Woodford and Harriss when they introduced them. The name Alisitos has priority (Santillán and Barrera, 1930) over San Telmo.

Figure 11. Northern end of the Sierra Pinta. Oblique aerial view to the east across the northern end of the Sierra Pinta to the mud flats of the Colorado River Delta. Mexican Highway 5 from Mexicali to San Felipe crosses the view. The northern end of the Sierra Pinta is composed of Carboniferous metasedimentary rocks (black) and Miocene-Pliocene volcanic strata (gray). Along the crest of the metasedimentary ridge farthest to the right (A; above the sand slopes), Roland McEldowney discovered the first Paleozoic fossils to be found in place in the bed rock of the Peninsular Ranges. In the left center, B is the depositional contact between Carboniferous chert, limestone, and slate (to the left) and tilted volcanic strata (to the right). Photograph by John Shelton.

Figure 12. Arroyo Volcán crossing Paleozoic strata. Oblique aerial view southeast along the main gulf escarpment east of El Mármol (loc. 33). The nearly flat old erosion surface is in the right foreground. The flat-lying Tertiary volcanic rocks (A) were deposited on the stable peninsula surface and have become isolated by erosion. The strata dipping to the left (east) are metamorphosed Paleozoic sedimentary rocks. Arroyo Volcán crosses the center of the picture. Paleozoic(?) fossils have been collected near B. The white limestone (C) is partially replaced by barite. Photograph by John Shelton.

3
Mesozoic Thermal Event

It has long been recognized that a belt of "Mesozoic" batholiths extends from the Aleutian Peninsula more than 6,000 km to Central America (Daley, 1933). Prominent in this chain are the Coast Range batholith of southeastern Alaska and British Columbia, the Idaho batholith, the Sierra Nevada batholith, and the Peninsular Ranges batholith of southern California and Baja California. This belt is located in a somewhat broader belt of metamorphosed and folded rocks.

Metamorphism and batholithic emplacement are commonly considered to be of Mesozoic age. Some plutons, however, are at least as old as late Paleozoic, some are as young as Miocene, and the region is still thermally alive in many places.

METAMORPHIC FRAMEWORK

In the Peninsular Ranges batholith, only the areas of lowest grade regional metamorphism show contact metamorphic effects. In such areas, contact metamorphism produces micaceous minerals for a few tens of meters from the contact, but, even next to the intrusive bodies, the relict sedimentary or volcanic structures are easily recognized.

The regional metamorphism is of the type described by Miyashiro (1962) as Abukuma, which is believed to record a pressure-temperature domain intermediate between that of shallow contact metamorphism and classical Barrovian regional metamorphism.

Although large areas of the peninsula are underlain by metamorphic rocks, no extensive study of them has ever been undertaken. Only in the Sierra San Pedro Mártir (loc. 28 [localities in Chap. 3 are shown in Pl. 2]; Woodford and Harriss, 1938), the Sierra de los Cucapas (loc. 9; Barnard, 1968b), and the Sierra Pinta (loc. 16; McEldowney, App. 1) are these rocks well described. During the course of study, hundreds of prebatholithic rocks were collected. Near the west coast, most of these are volcanic or volcaniclastic rocks with only incipient metamorphism. About 120 thin sections were cut from rocks in which obvious recrystallization had taken place. Because of the areal distribution of metamorphic grades, most of these are metasedimentary rocks, particularly metasandstone and argillite. Slate, phyllite, schist, and gneiss are progressively encountered as the peninsula is crossed from west to east. Typically, mica-quartz rocks are found with oligoclase-andesine plagioclase. Many rocks contain garnet, cordierite, andalusite, both columnar and acicular sillimanite, and, in some areas, staurolite. The occurrences of these minerals do not in all cases correspond to the coarseness of recrystallization, to recognizable metamorphic zones, or to differences in parent rock. Neither kyanite nor chloritoid was found. We recognize four major metamorphic rock associations:

1. Rocks showing little or no metamorphism. These are believed to exist only outside of the zone of major batholithic emplacement.

2. Slate, phyllite, and associated low-grade metamorphic rocks. These are found within the boundary and roof zones of the major batholithic belt and are intruded only by small plutons.

3. Schist, amphibolite, and related medium-grade metamorphic rocks. Much of their premetamorphic fabric still exists, and they are intimately involved with the major plutons.

4. Rocks in which the premetamorphic fabric has been destroyed. Structural relations and cooling history suggested by K-Ar ratios imply that these have undergone deep and prolonged burial. Some record complex metamorphic histories.

Rocks of metamorphic associations 2 and 3 and those of 3 and 4 are found together in some places.

We infer that the relations among the metamorphic associations are similar to those between the superstructure and infrastructure as proposed by Wegman (Turner and Verhoogen, 1960). Associations 1 and 2 represent the rigid domain of the crust (the superstructure) during plutonic emplacement and regional metamorphism (Fig. 13); association 4 represents the plastic infrastructure; and association 3 represents a transitional zone. The infrastructure, then, consists of magma and metamorphic rocks that deform plastically; the infrastructure was intruded into the superstructure, which yielded by block faulting and dilation. Mesozoic volcanism was synchronous with intrusion. In places along the sides of the infrastructural "rise" is a zone of mylonitization that marks the boundary between associations 3 and 4. The western boundary of the infrastructural association is best observed along the road between the southwestern corner of Valle San Rafael (loc. 17) and Pino Solo. The eastern boundary of the infrastructural association is exposed along the foothills of the gulf escarpment west of Laguna Salada (loc. 19).

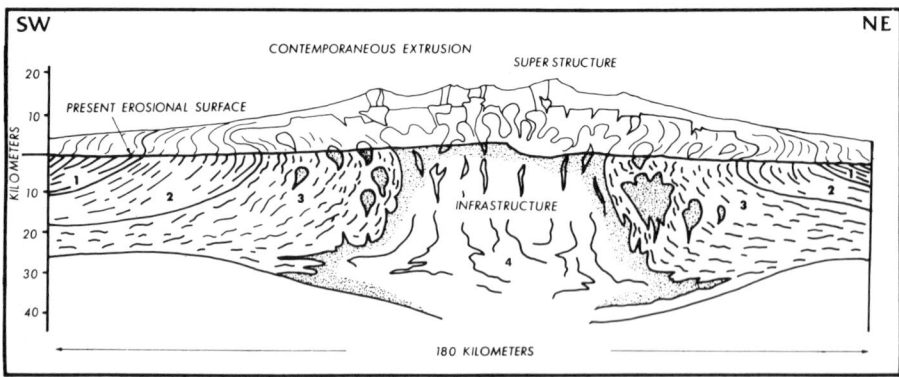

Figure 13. Mesozoic metamorphic-tectonic zones. A hypothetical cross section of the peninsula showing the relation of infrastructural and superstructural zones (numbers refer to description in text).

The present west coast of Baja California is very close to the western edge of the Mesozoic metamorphic belt. The eastern margin of the metamorphic belt is not far to the east of the gulf shore. This conclusion is based on the low grade of metamorphism of some of the Paleozoic rocks in the northern Sierra Pinta and the Cenozoic conglomerate containing unmetamorphosed Paleozoic limestone in the Sierra Santa Rosa (loc. 30). Yet the Mesozoic rocks on the Sonora coast are at least as metamorphosed as those on the eastern coast of Baja California. This suggests that (1) the Sonora coast rocks are not part of the Peninsular Ranges thermal belt, (2) the Sonora coast rocks display a deeper erosion level, or (3) large-scale strike-slip motion between the two coasts has duplicated the eastern margin. We favor the third alternative (Gastil and others, 1972).

PENINSULAR RANGES BATHOLITH

The first geologist to trace the extent of the Peninsular Ranges batholith was William M. Gabb (1882, p. 137). Lindgren (1888) was impressed by the enormous extent of the granitic rocks and the small areas occupied by metamorphic rocks. He recognized (1889) that the largest part of the batholith is quartz-mica diorite with a small percentage of orthoclase. This conclusion was confirmed by the work of Larsen (1948) north of the international border and by Silver and others (1963) south of the area visited by Lindgren.

Hirschi and de Quervain (1927-1933) published a number of petrographic descriptions and chemical analyses of plutonic rocks of Baja California. As their descriptions omit mineral percentages, it is impossible to compare their rock nomenclature with ours. Of 60 granitic rocks from the state of Baja California, they named 26 granodiorite, 12 quartz diorite or tonalite, 11 diorite or gabbro, 10 granite, and 1 syenite. Their chemical analyses suggest that their granite category includes what we would call adamellite and that their granodiorite category includes those tonalites that contain noticeable potassium feldspar.

Hirschi and de Quervain were the only authors to consider the granitic petrography of Baja California as a whole. Significantly, they observed (1933, p. 275) that potassium-rich granitic rocks are more abundant on the gulf coast and that noritic gabbro is more abundant near the Pacific coast.

Larsen (1948) described those portions of the Peninsular Ranges batholith in northern San Diego County and western Riverside County and saw an east-west zonation in composition of the plutonic rocks (p. 136-137):

Tonalites of several kinds and some related granodiorites low in potash feldspar make up nearly all the eastern half of the batholith. These rocks are lower in dark minerals and potash feldspar than the rocks of the western part of the batholith. They do not fall on the variation curves for the rocks of the western part of the batholith but are lower in K_2O, FeO, and MgO, and higher in SiO_2 and Al_2O_3. . . .The boundary between that part of the batholith in which granodiorite and gabbro are abundant and that in which they are lacking is about parallel to the structural trends of the region and to the elongation of the batholith.

The scattered bodies of granite in the desert ranges, east of the Peninsular Ranges, are chiefly granodiorite (Miller, 1946), and they fall near variation curves that are higher in K_2O, Na_2O, and Al_2O_3 and lower in SiO_2 than the rocks of the western part of the batholith. Thus, the part of the batholith studied appears to be divided into three parts, and the general character of the magmas of the three parts was different. The rocks of the western part of the batholith are moderate in K_2O, those of the eastern part of the Peninsular Ranges are low in K_2O, and those of the bordering area to the east of the Peninsular Ranges are high in K_2O. A detailed study of this feature of the batholith is desirable.

Hudson (1922), in describing the Cuyamaca Peak quadrangle, was the first of several authors to map and name plutonic map units within the Peninsular Ranges. Commonly, authors also indicated relative ages. In southern California, this was continued by Dudley (1935) in Riverside County, Miller (1935a) in southern San Diego County, Miller (1937) in west-central San Diego County, Merriam (1946) in the Ramona quadrangle, Larsen (1948) in northwestern San Diego and western Riverside Counties, Everhart (1951) in the Cuyamaca Peak quadrangle, and Merriam (1959) in the Santa Ysabel quadrangle.

Woodford and Harriss (1938) did the same in the Sierra San Pedro Mártir, Baja California (loc. 40, in Fig. 2). Several of the plutonic formations (namely, Dudley's Lakeview [now Lakeview Tonalite] and Miller's Woodson Mountain, Green Valley, Bonsall, and San Marcos) have been recognized and extended by subsequent authors.

With the exception of Hudson, who placed the Stonewall Quartz Diorite older than the gabbroic rocks of Cuyamaca Peak, each of the above authors believed that the plutonic succession followed a differentiation progression from gabbro to granodiorite or granite. Recent geochronologic work in the Peninsular Ranges batholith by Banks and Silver (1969) indicates that this assumed progression is not valid; rocks previously included within one formation have different ages, and granitic "formations" do not necessarily occur in the same sequence from area to area.

Distinct granitic types do appear again and again over wide areas. One such rock type, typified by Miller's (1935a) La Posta Quartz Diorite, occurs the entire length of the eastern half of the Peninsular Ranges province. This particular rock is distinctive for its idiomorphic biotite books, hornblende prisms, and the size and abundance of sphene crystals. Although the significance of distinctive "lithosomes" within the batholith is not as yet understood, the wide distribution of some and the symmetrical distribution of others do suggest that the batholith is a unit. The different parts may bear a time-and-space relation to one another and may have been generated by processes or events that were province-wide.

Recently, some representative plutons have been studied. Duffield (1968) described the El Pinal tonalite-granodiorite body in the west central Sierra Juárez (loc. 7; see also Fig. 14); Birkhahn (App. 2) described the San José tonalite pluton (loc. 22), just west of the Sierra San Pedro Mártir; and Itson (App. 1) described a body of gabbro near Jamul, a few kilometers north of the international border. A pluton just south of El Pinal (loc. 8) was mapped by Raymond Elliot (1968, written commun.), and the San José pluton was mapped in more detail (Smith and others, 1971; Murray, 1975).

Mineral Assemblages

The granitic rocks of the state of Baja California were mapped according to rock type, which was determined by hand lens and stain tests applied in the field. The identifications were checked by 306 rock samples stained in the laboratory and studied in thin section (Table 2). Plagioclase composition was determined by universal stage for 57 samples. Nine were chemically analyzed. Figure 15 shows the approximate proportions of quartz, potassium feldspar, and plagioclase for thin-sectioned rocks. Most of the analyses plot close to a line that begins at the 100 percent plagioclase corner (quartz-free gabbro and diorite) and extends along the plagioclase-quartz side (passing through the quartz gabbro-quartz diorite field to the tonalite field) to about the 60 percent plagioclase intercept. The line then turns sharply toward the quartz-potassium feldspar boundary and crosses the field of granodiorite and adamellite. Contours of equal anorthite content may be drawn

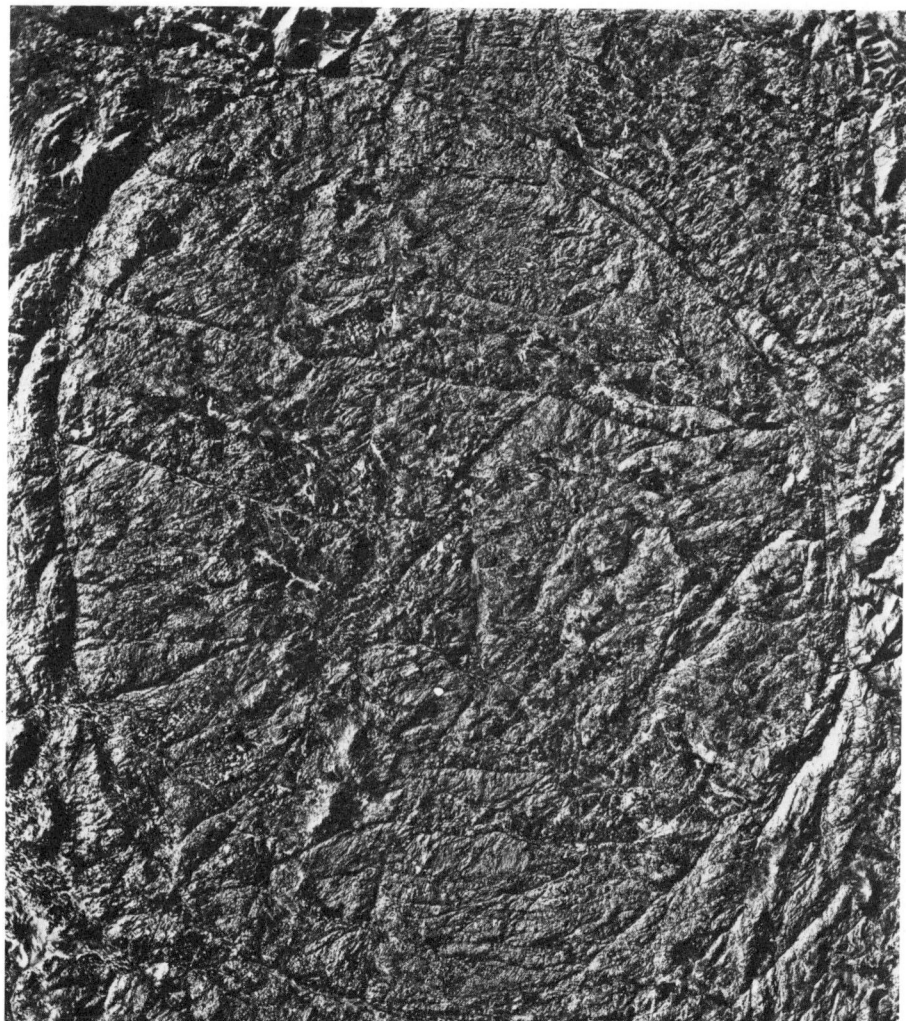

Figure 14. Vertical aerial view of the El Pinal pluton, 45 km south of the international border (north is at the top). The prominent circular joint pattern has a diameter of about 7 km. Although the foliation of the pluton conforms to the circular pattern that is apparent in the air photograph, the outer contact of the pluton is found outside the visual boundary; but this is not usual for plutons of the state of Baja California. The pluton has an inner and outer zone of tonalite with a medial ring of granodiorite. It intrudes amphibolite-grade metasedimentary rocks and older plutons. The topography is typical of the Tecate surface (see Chap. 8). The highlands are related to the Eocene or older erosion surface; the arroyos are incised as much as 250 m along the joint systems.

TABLE 2. THIN-SECTION IDENTIFICATION
OF GRANITIC ROCKS

Rock type	Number of samples	Percent of total
Granite	11	4
Adamellite	28	9
Granodiorite	52	17
Tonalite with K - feldspar	69	
Tonalite without K - feldspar	110	60
Quartz diorite	6	
Quartz gabbro	2	
Quartz norite	5	
Gabbro	6	10
Diorite	15	
Norite	1	
Anorthosite	1	
	306	

NOTE: The quartz — plagioclase — K - feldspar proportions are plotted on Figure 15.

TABLE 3. COMPOSITION OF PLUTONS

Predominant Rock Type	Plutons Number	%	Area Km^2	%
Gabbro and diorite	34	14	371	2
Tonalite and quartz diorite	118	47	11,518	73
Granodiorite	87	35	4,911	23
Adamellite	6	2	402	2
Granite	5	2	42	0.2
	250		17,244	

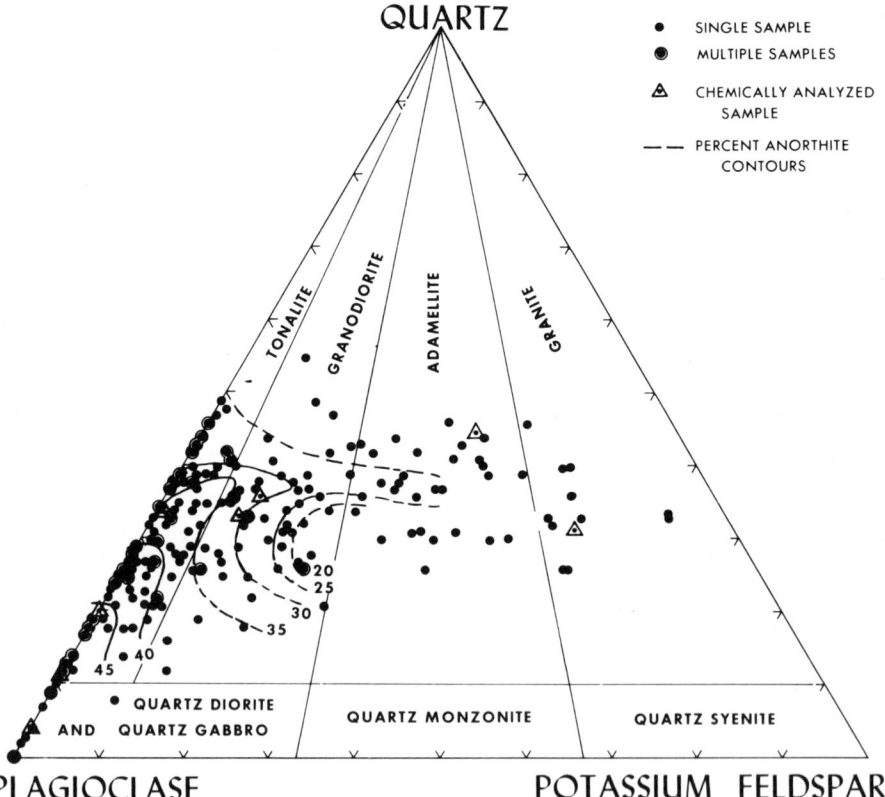

Figure 15. Quartz-potassium feldspar-plagioclase diagram for point-counted granitic rocks from the state of Baja California.

through the gabbro-diorite and tonalite fields but become complicated in the fields of adamellite and granite.

Figure 15 also shows the classification used for plutonic igneous rocks in this study. It is based entirely on the percentages of quartz, potassium feldspar, and plagioclase with the gabbro-diorite distinction being made at 50 percent anorthite. Field nomenclature was assisted by kits for staining potassium feldspar. Field distinction between gabbro and diorite was predicated on the identification of ferromagnesian minerals: olivine or pyroxene indicated gabbro; biotite indicated diorite; hornblende alone was indeterminate.

The names tonalite and quartz diorite are used synonymously in current literature. Very few authors use both names. Hirschi and de Quervain (1927-1933) did use both names for the rocks of Baja California but failed to define the difference. We use the term "quartz diorite," parallel to the sense of "quartz syenite" or "quartz gabbro," to indicate a rock with a very small amount of quartz (<10 percent; see Fig. 15). We use "tonalite," parallel to the sense of "granite" and "granodiorite," to indicate a rock with appreciable quartz (>10 percent). The term "quartz monzonite" (<10 percent quartz) is distinguished from "adamellite" in the same way.

In the state of Baja California, 387 separate plutons with average diameters of one kilometer or more have been recognized during reconnaissance mapping. Many plutons now believed to be a single body will undoubtedly be subdivided by more detailed mapping. The analysis is also complicated by the fact that many plutons vary in composition; 250 of the plutons have been identified according to their predominant rock composition (Table 3). This approximation of the composition of the batholith confirms the earlier estimates of both Lindgren and Silver.

Chemical Analysis of Batholithic Rocks

A sequence of nine rocks was selected for chemical analysis (Table 4; App. 4), beginning with one diorite and one gabbro, which are plotted at the plagioclase corner of Figure 15, and proceeding along the plotted trend of the samples.

Norms (Table 5) were obtained by a computer using the Barth formula (courtesy of James W. Hawkins, Univ. California, San Diego); they disagree with the modes. Contrary to the definitions based on modes and given in the previous section, the norms predict more than 10 percent quartz in the quartz gabbro and the quartz diorite and much more potassium feldspar in the diorite and tonalite than is identified under the petrographic microscope (Table 6).

The ratio K_2O/SiO_2 for the nine selected samples (App. 5) is almost as variable as that for all analyses of the Peninsular Ranges batholith. The nine analyzed rocks can be grouped in two populations: five with a high potassium content relative to silica and four relatively low in potassium (B3T-40, B3J-60, B5G-54, and B3M-21). When the variation diagrams for the other common oxides are compared to these two populations, each oxide except Na_2O and MgO shows a consistent relation to potassium content. In the potassium-rich rocks, potassium abundance is inverse to that of most major oxides but positive in reference to phosphorus oxide and alumina. The titania/silica ratios are also consistently higher in the high-potassium rocks, but the difference is too small to be significant. All five rocks with a high K_2O/SiO_2 ratio (high-index rocks, see Fig. 16) showed a marked norm deficiency of potassium feldspar (Table 6), whereas the rocks with a low K_2O/SiO_2 ratio (low-index rocks) do not.

The plots of K_2O/SiO_2 and $(K_2O + Na_2O)/SiO_2$ (Fig. 16) for all the analyses of Larsen (1948), Hirschi and de Quervain (1927-1933), and the new analyses from Baja California show that most of the rocks are confined to a rather well-defined trend, typical of a petrogenically related province. The median line of the trend falls along the boundary between Kuno's pigeonitic and high-alumina series (Kuno, 1969, p. 14), along the low $(K_2O + Na_2O)$ edge of the Cascade rocks (Kuno, 1969, p. 14), and a little lower than the east-central Sierra Nevada batholith (Hamilton, 1969a, p. 179). All but one of the gabbro samples fall almost exactly on the plot of Hawaiian tholeiite (Kuno, 1969, p. 18). The Mesozoic Peninsular Ranges batholith is considerably less alkalic than the Cenozoic volcanic rocks that overlie it (see Chap. 5) and is similar in composition to the Mesozoic volcanic rocks that it intrudes (see Chap. 2).

Larsen (1948) and Hirschi and de Quervain (1933) suggested that the batholith becomes more alkalic to the east. Hirschi and de Quervain based this on a single reconnaissance of the peninsula; Larsen's conclusion was based on comparing his own extensive petrographic and chemical data along the western edge of the batholith with the petrographic reconnaissance of others to the east. Moore (1959) generalized similar observations to define the "quartz diorite line," and Dickinson and Hatherton (1967) observed an increase in the K_2O/SiO_2 index from the oceanic to the continental side of igneous-orogenic belts.

We have been unable to verify these conclusions. A $(K_2O + Na_2O)/SiO_2$ index

TABLE 4. CHEMICAL ANALYSES OF PLUTONIC ROCKS FROM THE STATE OF BAJA CALIFORNIA

Sample	SiO_2	Al_2O_3	Fe_2O_3	FeO	MgO	CaO	Na_2O	K_2O	TiO_2	MnO	P_2O_5	H_2O	Total
B3T-40 *	50.69	17.76	5.40	5.58	5.19	10.58	2.15	0.20	1.00	0.18	0.10	0.82	99.65
B5F-48C *	50.42	18.48	3.42	4.75	4.82	9.04	3.29	1.33	1.08	0.15	0.28	2.26	99.32
B5G-54 *	56.42	16.50	5.45	3.99	3.31	7.73	4.49	0.20	1.00	0.17	0.16	0.75	100.17
B3M-35 *	56.67	16.77	4.06	3.70	3.67	7.19	3.47	1.14	1.09	0.12	0.21	1.17	99.26
B3K-41 *	62.24	16.33	2.76	3.16	2.87	5.34	3.07	2.42	0.79	0.10	0.16	1.41	100.65
B5F-2D *	70.11	15.44	0.75	0.98	0.94	2.83	4.42	2.73	0.35	0.04	0.11	0.80	99.50
B3J-60 *	73.38	13.93	0.82	0.79	0.44	2.49	4.18	2.34	0.25	0.03	0.07	0.66	99.38
B5E-53A *	73.49	14.17	0.73	0.33	0.45	1.44	3.20	5.25	0.30	0.03	0.10	0.59	100.08
B3M-21 *	77.50	13.02	0.36	0.62	0.10	1.24	4.18	3.78	0.10	0.03	0.04	0.51	101.48
2 †	64.66	14.97	1.41	4.43	2.24	5.02	3.46	1.91	1.23	0.13	0.27	0.48	100.21
3 †	64.60	16.12	1.59	3.45	2.23	5.12	3.02	2.32	0.69	0.07	0.22	0.67	100.10
4 †	66.65	14.70	1.51	3.83	1.55	4.43	3.32	2.75	0.67	0.09	0.29	0.41	100.20
12 †	62.35	15.28	2.02	5.10	2.66	5.37	2.37	2.95	0.81	0.11	0.34	0.56	99.92
17 †	71.21	13.91	1.27	2.24	0.50	1.86	3.08	4.96	0.34	0.04	0.28	0.45	100.14
23 †	62.44	15.20	1.39	5.10	2.67	5.22	2.80	2.92	0.87	0.05	0.21	0.84	99.71
24 †	57.22	16.76	1.13	6.53	4.14	7.40	2.26	1.43	1.16	0.11	0.38	1.08	99.60
25 †	71.42	11.93	2.15	3.10	0.85	3.21	2.94	3.05	0.58	0.05	0.21	0.55	100.04
27 †	53.05	18.89	3.63	5.79	4.17	8.38	2.74	0.87	1.62	0.10	0.28	0.70	100.22
34 †	49.70	19.77	1.72	7.57	6.06	11.28	2.10	0.48	0.59	0.11	0.10	0.85	100.33
36 †	50.03	20.96	2.85	5.16	5.70	12.09	2.24	0.25	0.42	0.12	0.06	0.64	100.52
41 †	72.96	13.32	1.06	2.54	0.46	1.83	3.05	3.88	0.40	0.03	0.07	0.37	99.97
50 †	66.50	16.08	0.85	4.05	1.28	3.93	3.21	2.41	0.64	0.04	0.15	0.87	100.01
61 †	64.60	16.49	1.09	4.14	1.80	4.14	4.40	2.06	0.63	0.03	0.17	0.57	100.12
64 †	71.42	14.01	0.81	2.64	0.87	2.00	3.26	3.73	0.52	0.03	0.26	0.56	100.11

* See Plate 1-A, 1-B, and 1-C for locations and Appendix 4 for petrographic descriptions.
† Hirschi and de Quervain, 1927 - 33.

TABLE 5. MODES AND NORMS FOR PLUTONIC ROCKS

Minerals or normative molecules present, in %

Sample	Type of determination*	Plagioclase	Albite	Anorthite	K - Feldspar	Quartz	Biotite	Hornblende	Augite	Hypersthene	Opaques	Others	Secondary	Ilmenite	Iron Oxides	Ferrosilite	Enstatite	Others
B3J-60	e	42	..	..	18	35	3	0	0	0	T	M Z A S	C E	..	..	..	..	..
	pc	49	..	..	10	33	5	0	0	0	0	..	..	..	..	..	..	..
	n	50.4	38.3	12.1	14.1	32.3	..	..	..	..	..	0.1A	..	0.4	0.9m	0.3	1.2	Co
B3M-21	e	15	..	..	47	35	3	0	0	0	T	2A	C	..	..	..	..	..
	pc	18	..	..	49	30	3	0	0	0	T	..	..	..	..	..	..	..
	n	43.1	37.5	5.6	22.3	33.1	..	..	..	..	..	0.8A	..	0.1	0.4m	0.6	0.3	W
B3M-35	e	63	..	..	0	17	6	11	3	0	T	Z A S	Se C E	..	..	..	..	..
	pc	59	..	..	0	16	8	13	3	0	T	..	..	..	..	..	..	..
	n	59.6	32.0	27.6	6.9	12.1	..	..	..	..	..	0.5A	..	1.6	4.4m	1.4	10.4	W
B3T-40	e	68	..	..	0	T	T	18	1	10	2	Z A S T	..	..	..	..	..	..
	pc	60	..	..	0	3	T	22	T	10	3	..	..	..	..	..	..	..
	n	59.3	19.9	39.4	1.2	7.9	..	..	..	..	..	0.2A	..	1.4	5.8m	3.6	14.8	W
B3K-41	e	53	..	..	3	30	7	6	1	0	T	A	Se E C	..	..	..	..	..
	pc	41	..	..	1	31	13	11	T	0	T	..	..	..	..	..	..	..
	n	52.2	28.1	24.1	14.6	18.1	..	..	..	..	..	0.3A	..	1.1	2.9m	1.9	8.1	W
B5G-54	e	75	..	..	0	7	1	6	1	9	1	A S	C	..	..	..	..	..
	pc	70	..	..	0	8	0	11	2	8	4	..	..	..	..	..	..	..
	n	65.5	40.9	24.6	1.2	10.1	..	..	..	..	..	0.4A	..	1.4	5.8m	1.0	9.3	W
B5F-48C	e	68	..	..	0	1	4	16	0	0	T	Z A S	Se E C	..	..	..	..	..
	pc	58	..	..	0	2	7	22	0	0	T	..	..	..	..	..	..	..
	n	63.1	30.4	32.7	8.1	0.9	..	..	..	..	..	0.6A	..	1.5	3.7m	3.6	13.7	W
B5F-2D	e	50	..	..	10	35	4	1	0	0	0	Z A	C E	..	..	..	..	..
	pc	51	..	..	9	29	9	T	0	0	0	..	..	..	..	..	..	..
	n	53.7	40.2	13.5	16.3	24.9	..	..	..	..	..	0.2A	..	0.5	0.8m	0.5	2.6	Co
B5E-53A	e	27	..	..	35	35	3	0	0	0	T	M Z A	C	..	..	..	..	..
	pc	23	..	..	32	41	3	0	0	0	T	..	..	..	..	..	..	..
	n	35.7	29.1	6.6	31.5	29.4	..	..	..	..	..	0.2A	..	0.4	0.4H 0.1m	0.0	1.3	Co

T - trace   A - apatite   Se - sericite   Co - corundum
M - muscovite   S - sphene   E - epidote   W - wollastonite
Z - zircon   C - chlorite   H - hematite   m - magnetite

* e - estimate; pc - point-count mode; n - norm.

TABLE 6.
COMPARISON OF POINT-COUNT MODES AND NORMS FOR THE FELSIC COMPONENTS OF PLUTONIC ROCKS

Sample No.	SiO_2* %	Type of Determination†	As % of total felsic component			Anorthite as % of Plagioclase	Rock name ‡
			Quartz	K-feldspar	Plagioclase		
B3T-40	50.69	pc	5	0	95	63 - 75	Quartz Gabbro
		n	12	2	87	66	Quartz Gabbro
B5F-48C	50.42	pc	3	0	97	39 - 47	Quartz Diorite
		n	1	11	88	52	Gabbro
B5G-54	56.42	pc	10	0	90	41 - 47	Tonalite
		n	13	2	85	38	Tonalite
B3M-35	56.67	pc	21	0	79	34 - 47	Tonalite
		n	15	9	76	46	Tonalite
B3K-41	62.24	pc	42	1	56	30 - 39	Tonalite
		n	21	17	61	46	Granodiorite
B5F-02D	70.11	pc	33	10	57	23 - 36	Granodiorite
		n	29	19	51	25	Granodiorite
B3J-60	73.38	pc	36	11	53	19 - 38	Granodiorite
		n	33	15	52	24	Granodiorite
B5E-53A	73.49	pc	43	33	24	14 - 38	Adamellite
		n	30	33	37	18	Adamellite
B3M-21	77.50	pc	31	51	19	14 - 20	Granite
		n	34	22	44	13	Adamellite

* From Table 4.
† pc - point-count mode; n - norm.
‡ See Figure 15.

was assigned to each analysis, based on zones drawn parallel to the main trend lines on the $(K_2O + Na_2O)/SiO_2$ variation diagram (Fig. 16A). When these indexes are plotted relative to an arbitrary median line drawn down the axis of the batholith from the Transverse Ranges to lat 28° N., no systematic increase in $(K_2O + Na_2O)$ to the northeast is revealed (Fig. 16B). The poor areal distribution of points renders the analysis inconclusive; with a little imagination, however, the diagram can be interpreted as supporting Larsen's conclusion (1948, p. 136-137) that the batholith is divided into three belts with the central belt being poorest in K_2O and Na_2O. A similar analysis using only K_2O is even less conclusive (Fig. 16C).

Sizes of Plutons

In the state of Baja California, 387 plutons have average diameters greater than 1 km; these plutons cover a total area of almost 28,000 km². Seven large plutons mapped as single bodies account for a third of this total area. The most common exposure diameter is 4 km (Fig. 17). Gabbro and granite form the smallest plutons with exposure diameters smaller than 2 km. The granodiorite-adamellite bodies range widely in size. Most of the largest plutons are tonalite. These relations are like those found in the Sierra Nevada batholith (Gastil and others, 1971). Most of the individual plutons (Pl. 2) are found in the marginal portions of the batholith, whereas the axial portions are occupied by a relatively small number of much larger bodies.

Many of the pluton exposures show a regular circular outline (locs. 7, 20, and 33; see also Fig. 18). If it is assumed that most of the bodies are diapirs and tend toward an ideal teardrop shape, plutons of a given size will yield a variety of diameters depending on the level to which each pluton is eroded. If all erosion depths are equally likely, and the diapiric shape is assumed to be an inverted teardrop, then the pluton exposure will show a preferred diameter approximately two-thirds of their maximum diameter. In addition, those diapirs showing their

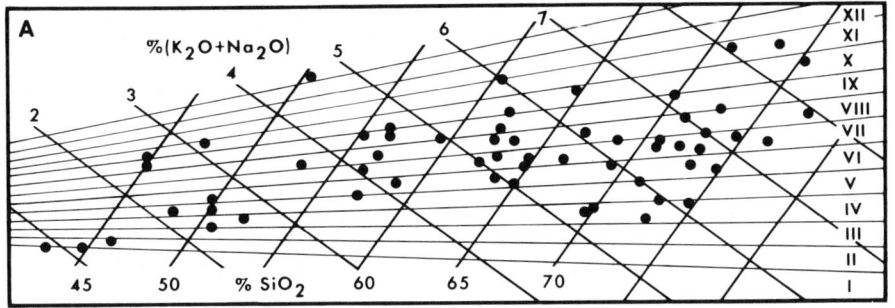

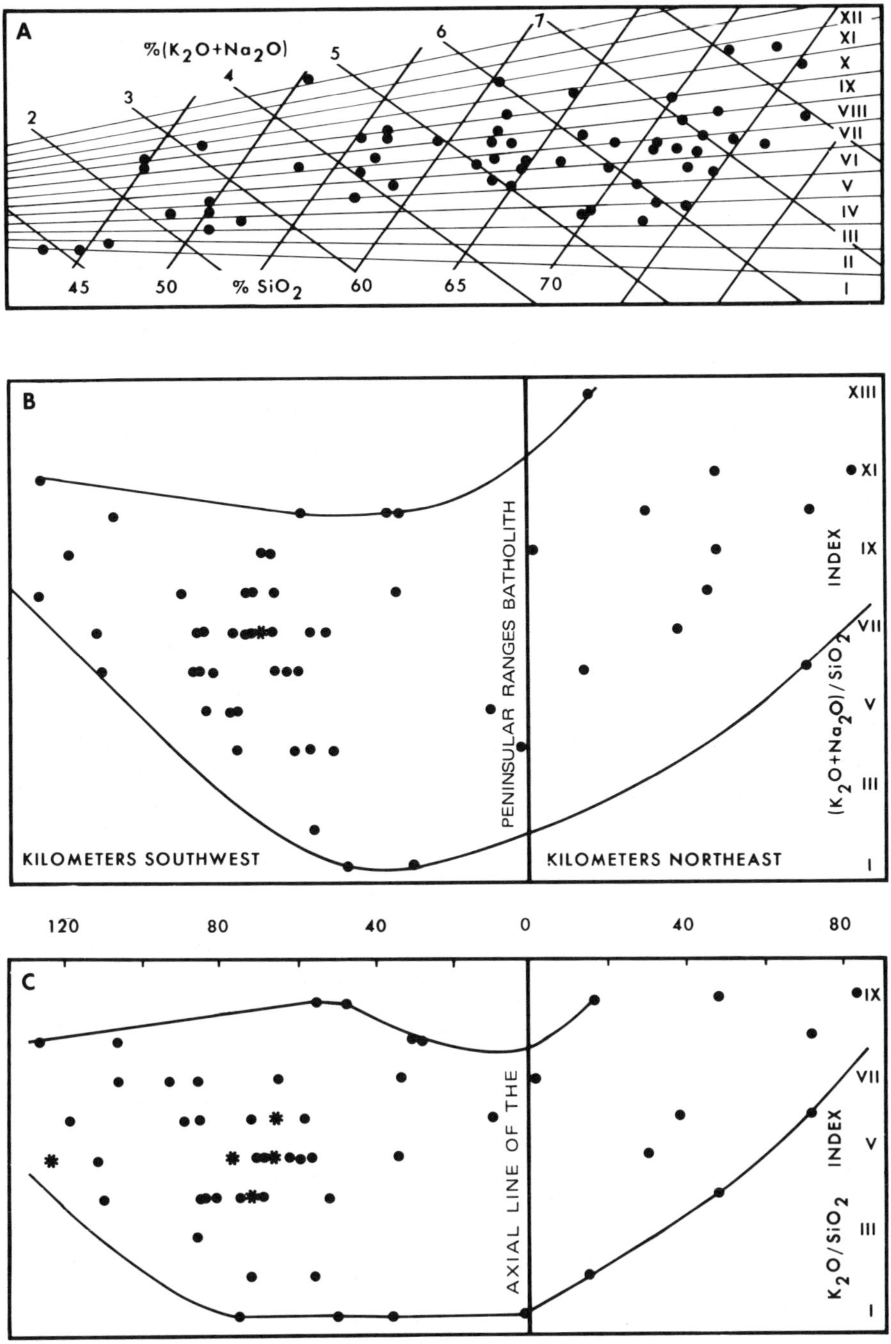

Figure 16. (A) Derivation of $(K_2O + Na_2O)/SiO_2$ index (roman numerals) from $(K_2O + Na_2O)/SiO_2$ variation diagram. (B) Plot of $(K_2O + Na_2O)/SiO_2$ index against the relative distance southwest or northeast from an arbitrary axial line through the Peninsular Ranges batholith. (C) A similar plot for K_2O index only.

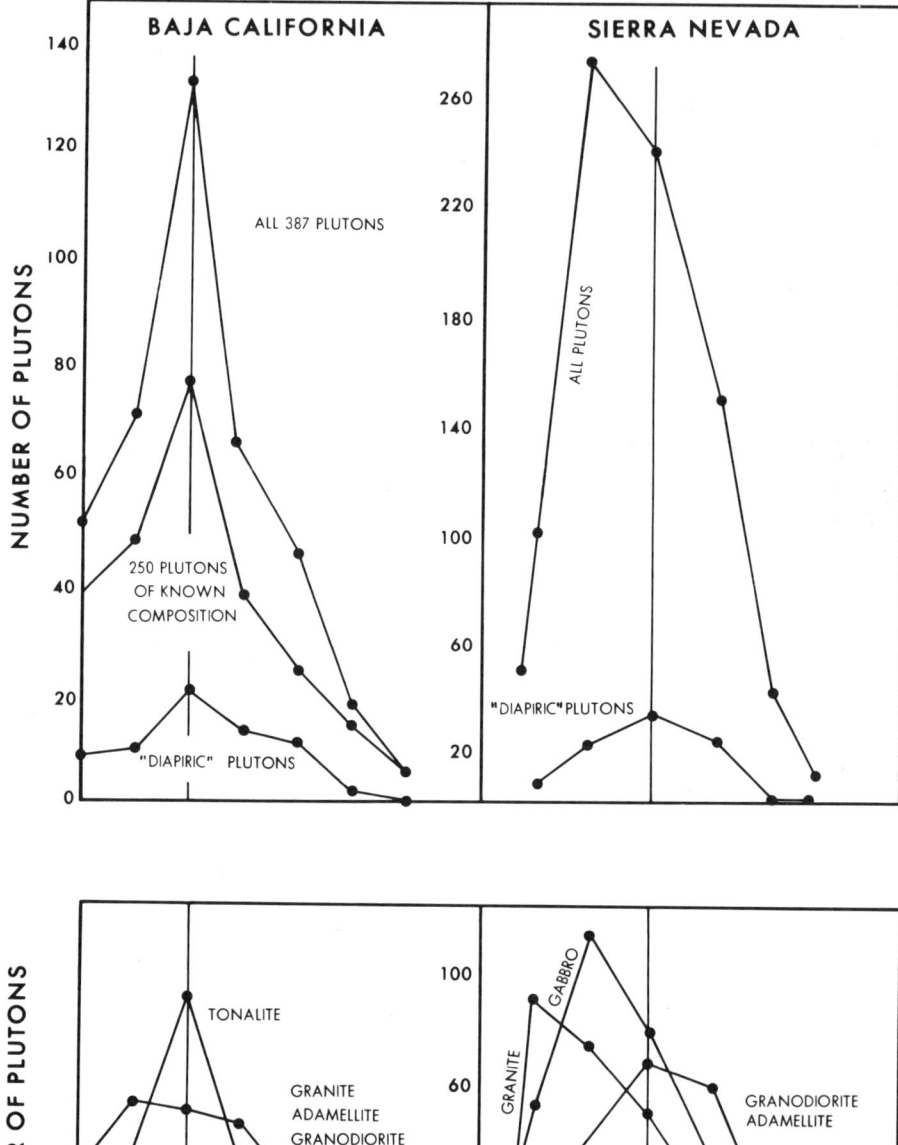

Figure 17. Size distribution of plutons of the state of Baja California compared with the size distribution of plutons of the Sierra Nevada, California, U.S.A., according to composition and shape. "Diapiric" refers to plutons with exposures similar to those in Figure 18. (Sierra Nevada distribution from Gastil and others, 1971.)

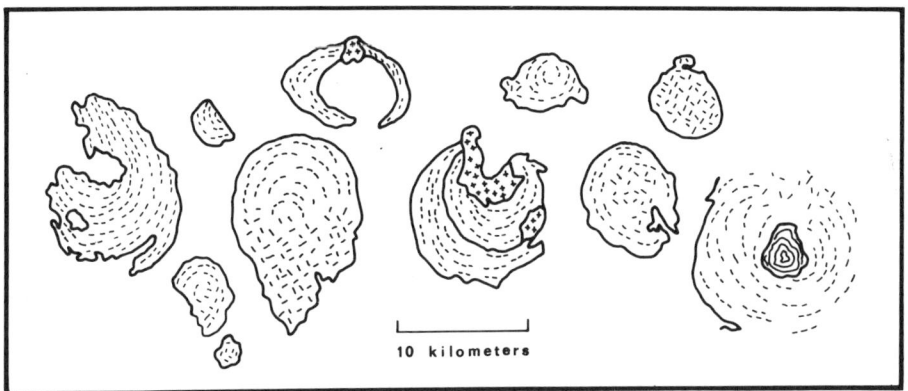

Figure 18. Shapes of some concentrically structured plutons in the state of Baja California.

maximum diameter would be most likely to have smooth-edged circular or ring-shaped cross sections. A size-distribution plot of the plutons that exhibit such a circular "diapiric" shape (Fig. 17) shows that the average diameter is very similar to (slightly larger than) the average diameter of all 387 plutons. If we are correct that these shapes indicate a diapiric intrusion process and if the preferred size is related to that process, then the similarity in preferred size to that of nondiapiric-shaped plutons suggests that the majority of the plutons have had their size (volume) determined by the diapiric process and have subsequently lost their concentric shape. Diapiric exposure shapes are found among plutons of all compositions.

Although later mapping may show that some of the plutons along the axis of the peninsula are composite, we believe that most of these bodies are continuous beneath Cenozoic cover and are relatively homogeneous. The 4-km peak in Figure 17 represents the statistically most likely exposure diameter for superstructure diapirs. An average exposure diameter of 4 km, assuming a teardrop shape, indicates diapiric bodies with a maximum diameter of 6 km. If these relatively small bodies were able to penetrate the cool superstructure, the giant infrastructural plutons may have literally lifted and broken the crustal roof to pour out large volumes of andesite and dacite (Fig. 13).

Concentric layering is found in many plutons when they are studied in detail (Duffield, 1968; loc. 7). Layering is particularly important in gabbro-bearing plutons. Examples are a gabbro to gabbroic anorthosite body east of Rancho San Pablo (loc. 15) and a large norite-cored pluton south of Valle Trinidad (loc. 20). The layered gabbro bodies have crystal cumulate structures, suggesting that these diapirs were derived from horizontally layered rocks.

Ages of Plutons

Some of the plutons clearly intrude the Alisitos Formation, which is of Aptian-Albian age based on fossil evidence (Allison, 1964). Silver and others (1969) dated a volcanic rock in the Alisitos Formation near Punta Cabra (loc. 12) at 127 ± 5 m.y. B.P. Field evidence, however, indicates that granitic rocks may have been intruded contemporaneously with and even earlier than the deposition of some of the Mesozoic volcanic rocks.

In San Diego County, M. A. Hanna (1926) reported granitic clasts in a volcaniclastic breccia that probably correlates with the rocks Fife and others (1967) dated as

Late Jurassic. West of Vallecitos (loc. 3), Ashley (App. 2) discovered metamorphosed volcanic and volcaniclastic deposits on an eroded granodiorite surface. The dike swarms that cut many of the plutons may have been feeders to the Mesozoic volcanic rocks. Snyder (App. 2) and Ashley (App. 2) described the relations near Vallecitos (loc. 4) that are illustrated diagrammatically in Figure 19. Furthermore, the close chemical similarity between the Mesozoic volcanic and plutonic rocks (Hawkins, 1970b) suggests that they were products of the same event.

Silver and others (1969) and Banks and Silver (1969) reported Pb-U isotopic dates ranging from 102 to 119 m.y. for granitic rocks in the western half of the Peninsular Ranges of southern California and Baja California. Silver (1969, oral commun.) subsequently determined ages as young as 95 m.y. for rocks in the eastern half of the peninsula. Field relations suggest that intrusion was in part contemporaneous with the deposition of the prebatholithic volcanic rocks. The isotopic dates indicate that some plutonism occurred well after the deposition of the youngest recognized prebatholithic strata. The plutons cut by the dike swarms may be older than those granitic rocks that have so far been dated.

Ages of Metamorphism and Cooling

Twenty-seven K-Ar ages (indicating the closing of the K-Ar system) have been determined for batholithic and prebatholithic rocks in the Peninsular Ranges of Baja California. Two of these were reported by McFall (1968) for the Bahía Concepción area and one pair by Duffield (1970, written commun.) for the El Pinal pluton (loc. 7). The others (Table 7) have been determined by Daniel Krummenacher and students at San Diego State University. The oldest of these is 107 m.y. for a volcanic horizon in the pre-Alisitos Jurassic(?) strata near Arroyo San José (loc. 32). These strata are weakly metamorphosed, and the data may represent the approximate age of metamorphism. In the western part of the peninsula where individual plutons intrude upper greenschist- or amphibolite-facies rocks, the ages are 90 m.y. or older. In the axial region of the peninsula where foliated plutons and sillimanite-grade gneiss are common, K-Ar ages are from 75 to 85 m.y., whereas in the desert ranges facing the Gulf of California, the K-Ar ages are from 60 to 69 m.y.

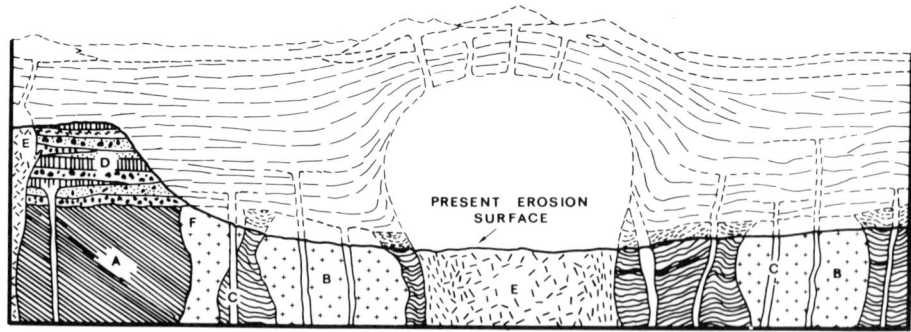

Figure 19. Diagrammatic sketch of relations between dike swarms and plutons near Vallecitos (loc. 4 in Fig. 4). A, slate and wacke host rock; B, predike plutons; C, swarm of primarily dacite dikes; D, volcanic-volcaniclastic host rock, believed to be the extrusive equivalent of the dacite dikes; E, postdike plutons; F, point where unit D rests directly on unit B.

This apparent decrease in K-Ar ages from west to east can be interpreted in several ways. We doubt that there was widespread plutonic emplacement and metamorphism as recently as 80 m.y. ago and certainly not 60 to 65 m.y. ago. The explanation may lie in a history of long-delayed cooling (Krummenacher and others, 1975). If the closing of the rocks to argon coincided with uplift and erosional unroofing, then the western part of the peninsula was uplifted first and was followed in succession by the more easterly portions. Such uplift is consistent with what we know of Upper Cretaceous and early Tertiary sediment provenance. Seven K-Ar ages were reported by Evernden and Kistler (1970) for the Peninsular Ranges of southern California. These ages are mainly from the western part of the province and fall between 86 and 105 m.y. Additional K-Ar determinations (Armstrong and Suppe, 1973) extended this work eastward and showed the same younger-to-the-east pattern found south of the international border.

TABLE 7.
K-Ar "AGES" FOR BATHOLITHIC AND PREBATHOLITHIC ROCKS
IN THE STATE OF BAJA CALIFORNIA

Locality*	Sample Nos.	K-Ar Age† in m.y.	Mineral	Rock and Locality
32	570	107.5 ± 3.0	Plagioclase	Pre-Alisitos volcaniclastic strata just north of the mouth of Arroyo San José, Punta Canoas quadrangle
32	537	89.1 ± 3.0	Biotite	Basic dike cutting 570. Both rocks have been weakly metamorphosed
14	573	94.5 ± 3.0	Hornblende	Metamorphosed diabase dike, southeast of El Alamo, Santa Catarina quadrangle
14	572H	83.7 ± 2.5	Hornblende	Pegmatite dike, cut by 573
31	545	91.9 ± 2.8	Biotite	North-south dike cutting tonalite southwest of La Virgen, El Mármol quadrangle. The dike has undergone amphibolite-grade metamorphism
13	511	92.1 ± 2.8	Biotite	Predike tonalite near Rancho Santa Clara, Ensenada quadrangle. Tonalite has undergone low-grade metamorphism
6	602	95.9 ± 2.9	Plagioclase	Dacite dike southeast of El Burro, San Pedro quadrangle. Cuts adamellite. Dikes appear unmetamorphosed
5	605	95.5 ± 1.3	Plagioclase	A different dike of the same dike swarm as 602
4	600	96.4 ± 3.0	Hornblende	Large granodiorite-tonalite pluton, El Testerazo, San Pedro quadrangle, cuts dike swarm which includes 602 and 605
4	601	99.2 ± 3.1	Biotite	
29	519	80.9 ± 2.5	Biotite	Coarse lamprophyre dike cutting tonalite, southeast corner of Valle San Felipe quadrangle
26	521 (B8B-21)	78.7 ± 2.5	Biotite	Cerro Borrego tonalite, San Felipe quadrangle; youngest of intrusive sequence by cross-cutting relations
27	507	82.9 ± 2.0	Biotite	Sillimanite gneiss, the oldest rock in San Matías Pass, San José de Castillo quadrangle
27	568	80.0 ± 2.0	Hornblende	Hornblende schist, San Matías Pass
27	501	77.3 ± 2.0	Biotite	Metamorphosed pegmatite, cuts 507
27	509	81.4 ± 2.0	Biotite	Metamorphosed gabbro, cuts 507 and 501
27	503	80.8 ± 2.0	Biotite	Metamorphosed basic dike, cuts 507 and 501
27	569	80.0 ± 2.0	Biotite	Granodiorite (unmetamorphosed?) cuts 507, 501 and 503
27	567	79.2 ± 2.0	Biotite	Aplite dike, cuts 569
27	505	79.0 ± 2.0	Muscovite	Pegmatite dike associated with 567
1	514B	75.1 ± 2.5	Biotite	Tonalite from In-Ko-Pah Gorge, east of Jacumba just north of the international border in San Diego County, California
25	520	64.5 ± 1.5	Biotite	Adamellite from the Sierra Pinta, southeastern corner of the Condensado quadrangle; the oldest of the plutonic sequence by cross-cutting relations
20	515	62.6 ± 1.5	Biotite	Tonalite from La Puerta, Sierra de los Cucapas, Laguna Salada quadrangle (collected by Barnard, 1968b)

* See Plate 2.
† Determined by Daniel Krummenacher, K-Ar Laboratory, San Diego State University.

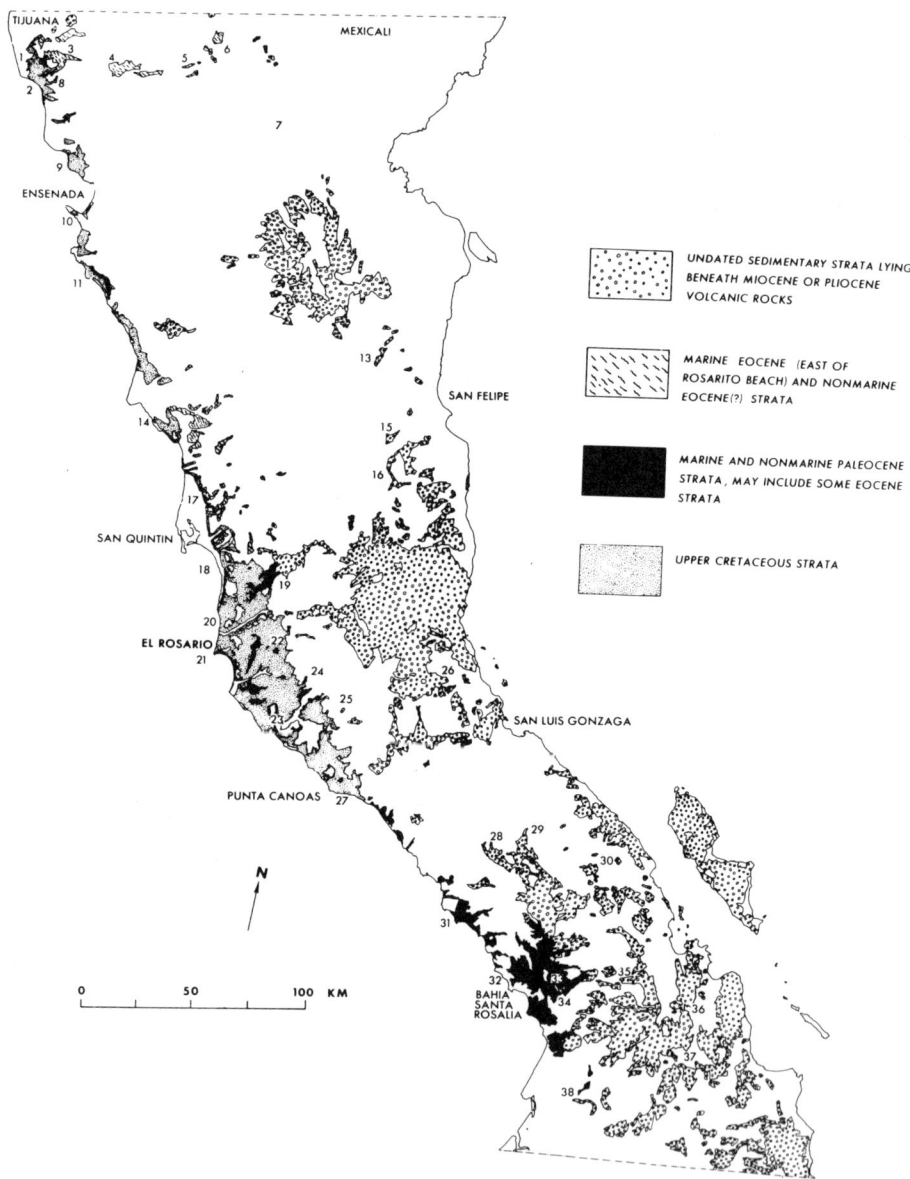

Figure 20. Distribution of Upper Cretaceous and lower Tertiary rocks. Numbers 1 through 39 correspond to localities cited in the text.

4
Late Cretaceous and Early Tertiary Time

Typical postorogenic activity occurred during Late Cretaceous through Paleocene and Eocene time. Following the emplacement of batholithic granitic rocks, probably completed about 90 m.y. ago, the Peninsular Ranges underwent uplift, cooling, and erosion. The principal areas of Mesozoic plutonic activity have remained emergent to this day.

A tremendous volume of eroded debris was transported westward to the Pacific Ocean and was deposited near the present shoreline. The outcrops of Upper Cretaceous and lower Tertiary rocks that resulted are shown in Figure 20 as well as the locations cited in Chapter 4. The Upper Cretaceous deposits were derived entirely from the granitic and metavolcanic rocks of the western and central parts of the peninsula. The resulting conglomeratic and arkosic debris built molasse-type deltas out to the edge of the continental slope.

By Eocene time, the Mesozoic mountains were reduced to isolated hills separated by alluviated pediplains across which flowed rivers that rose in the interior of the continent. Many of the clasts in the Eocene and Paleocene deposits were derived from rocks east of the central portion of the peninsula and were transported across the newly formed pediplains toward the Pacific Ocean.

Following Eocene time, the crustal areas now occupied by the Basin and Range province, including the Gulf of California, began to subside relative to the crustal area now occupied by the peninsula of Baja California. The rivers from the interior drained into interior basins and the proto-Gulf of California.

POSTBATHOLITHIC CRETACEOUS ROCKS

Stratigraphic Nomenclature

The first evidence for Upper Cretaceous rocks in Baja California was obtained by White (1885) in the vicinity of Punta Banda (loc. 10). Gabb (1882) included almost all the rocks of the peninsula that underlie Cenozoic volcanic strata in a unit that he called Mesa Sandstone, believed by him to be of Miocene age.

Lindgren (1888) speculated (p. 181-182) that Gabb's Mesa Sandstone included rocks of Late Cretaceous age. Emmons and Merrill (1894, p. 511), seeking to check this hypothesis, discovered Upper Cretaceous strata in the coastal area between Socorro (loc. 18) and Punta Canoas (loc. 27) but, like those who have followed, found no organic evidence to determine the time of deposition of the clastic strata that cap the interior of the peninsula.

In 1921 (p. 727), Darton described the coastal area as follows:

The upper part of the Chico Formation (late Cretaceous) rises above sea level a few miles north of Rosario, and it remains in view along the ocean bluffs and lower parts of the valleys of the Rio Rosario, Arroyo San Vicente, Rio San Fernando, and Arroyo Santa Catarina as far as latitude 29° 24', a few miles southeast of Punta Canoas. The rocks are soft sandstone and shale of light-gray to buff color with round concretions at most places.

The term "Chico" was used informally for the postbatholithic Cretaceous rocks until Beal (Marland Oil Company of Mexico, 1924) adopted the name Rosario for the strata exposed near the village of that name. Santillán and Barrera (1930) extended the name Rosario, applying it to the more northern, previously recognized Upper Cretaceous strata mentioned by White and Lindgren.

Anderson and Hanna (1935, p. 6) referred these strata to the "Chico Group" and, in addition to Beal's Rosario Formation, recognized a "Catarina Formation" with a type locality "about the lower part of the Arroyo Santa Catarina." They considered the strata at Arroyo Santa Caterina to be upper Senonian and the strata at El Rosario to be Turonian or older. The new formation name was predicated entirely on a supposed age difference interpreted from faunas.

Mina (1957) named Upper Cretaceous strata found along the Pacific coast in the lower half of Baja California Formación Valle Salitral (or Formación Valle) and correlated them with the Rosario Formation.

Kirk and MacIntyre (1951) referred the Upper Cretaceous of Punta San Isidro (loc. 11) to the Maestrichtian Stage, comparable in faunal aspect and in age to the "Chico Formation" of San Diego County, California, and other areas immediately north of the international border. Popenoe and others (1960) considered correlation of west coast Cretaceous rocks.

Kilmer's (1963) mapping of the area surrounding El Rosario resulted in a recognition of four formations of Late Cretaceous age (Fig. 21). In a printed abstract of the still unpublished work, Kilmer (1965) retained the name Rosario for only the upper marine formation. Names proposed for underlying formations, all parts of Beal's original Rosario Formation, were El Gallo Formation (for the underlying nonmarine unit), Punta Baja Formation (for a still lower marine unit), and La Bocana Roja Formation (for a basal nonmarine unit in the Upper Cretaceous sequence). The total measured thickness of the sequence near El Rosario is 2,900 m. The formation name El Gallo has been adopted in recent literature (Morris, 1966, 1967). Durham and Allison (1960) determined that the Rosario Formation (restricted definition introduced by Kilmer, 1965) straddles the Campanian-Maestrichtian boundary.

In recent years, the name Rosario Formation has been extended to include the rocks of San Diego County (Milow and Ennis, 1961; Holden, 1964; Sliter, 1968). Kennedy and Moore (1971) designated all the postbatholithic Cretaceous rocks of San Diego County as the Rosario Group, which they subdivided into the Cabrillo, Point Loma, and Lusardi Formations (Fig. 21).

The name "Catarina Formation," based entirely on an age difference that appears invalid, should be discarded.

The oldest postbatholithic Cretaceous strata so far identified in the state of Baja California are the Campanian rocks described by McGee (1965) at Punta Baja (loc. 21) south of El Rosario. He correlated the Punta Baja fauna with the

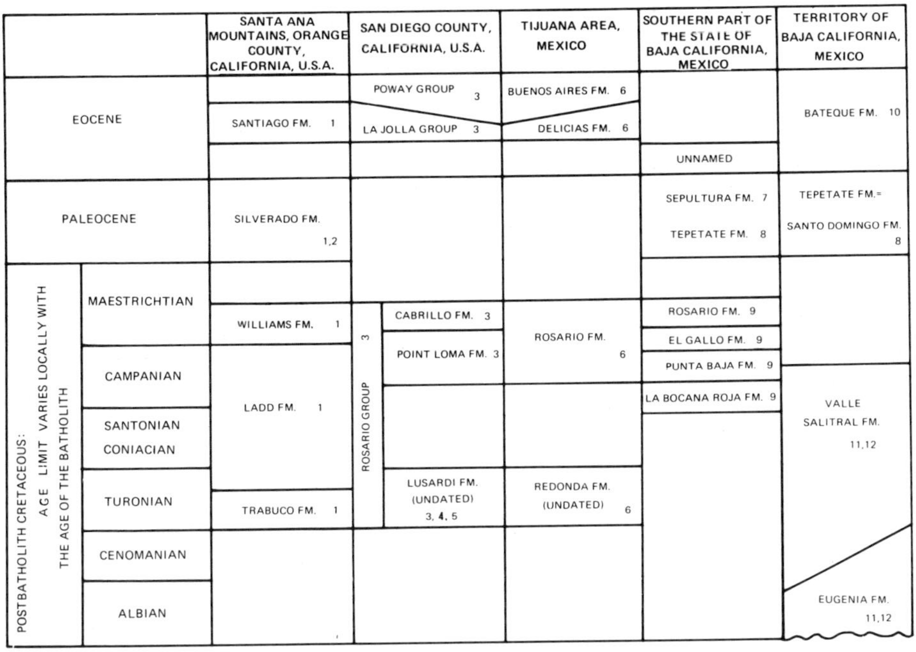

		SANTA ANA MOUNTAINS, ORANGE COUNTY, CALIFORNIA, U.S.A.	SAN DIEGO COUNTY, CALIFORNIA, U.S.A.	TIJUANA AREA, MEXICO	SOUTHERN PART OF THE STATE OF BAJA CALIFORNIA, MEXICO	TERRITORY OF BAJA CALIFORNIA, MEXICO
EOCENE		SANTIAGO FM. 1	POWAY GROUP 3	BUENOS AIRES FM. 6		BATEQUE FM. 10
			LA JOLLA GROUP 3	DELICIAS FM. 6		
					UNNAMED	
PALEOCENE		SILVERADO FM. 1,2			SEPULTURA FM. 7	TEPETATE FM.= SANTO DOMINGO FM. 8
					TEPETATE FM. 8	
POSTBATHOLITH CRETACEOUS: AGE LIMIT VARIES LOCALLY WITH THE AGE OF THE BATHOLITH	MAESTRICHTIAN	WILLIAMS FM. 1	CABRILLO FM. 3	ROSARIO FM. 6	ROSARIO FM. 9	
			POINT LOMA FM. 3		EL GALLO FM. 9	
	CAMPANIAN	LADD FM. 1	ROSARIO GROUP		PUNTA BAJA FM. 9	
	SANTONIAN				LA BOCANA ROJA FM. 9	VALLE SALITRAL FM. 11,12
	CONIACIAN					
	TURONIAN	TRABUCO FM. 1	LUSARDI FM. (UNDATED) 3, 4, 5	REDONDA FM. (UNDATED) 6		
	CENOMANIAN					
	ALBIAN					EUGENIA FM. 11,12

Figure 21. Correlation of postbatholithic Cretaceous, Paleocene, and Eocene formations. Numbers opposite formation names indicate the following sources: 1, Morton (1972); 2, Yerkes (1957); 3, M. A. Hanna (1926); 4, Kennedy and Moore (1971); 5, Nordstrom (1970); 6, Flynn (1970); 7, Santillán and Barrera (1930); 8, Heim (1922); 9, Kilmer (1963); 10, Mina (1957); 11, Popenoe and others (1960); 12, G. Gastil, J. Minch, and others (unpub. data).

fauna of Taylor age on the coast of the Gulf of Mexico. The base of this local section is unexposed.

Nordstrom (1970) described the undated conglomeratic Lusardi Formation in west-central San Diego County. He believed that it underlies the Rosario Formation and may correlate with the Trabuco Formation of the Santa Ana Mountains. Flynn (1970) recognized the analogous Redonda Formation in the area inland from Rosarito Beach (loc. 1). Both the Lusardi and Redonda Formations are characterized by very immature deposits of locally derived debris apparently deposited in a nonmarine environment. The upper surface of both is a deeply weathered, high-relief unconformity.

Sedimentary Rocks

In the past few years, a number of student research projects at San Diego State University have been concerned partially or wholly with Upper Cretaceous strata. Reed (App. 1), working east of El Rosario, found evidence of Cretaceous torrents that issued from steep-walled canyons hundreds of meters deep and carried angular boulders of locally derived granitic and metavolcanic rocks as much as 3 m in diameter. The lower red strata (Kilmer's La Bocana Roja Formation) consist mainly of metavolcanic debris. Overlying strata are white from the high percentage of granitic debris. Some beds consist of 75 percent granitic boulders in a matrix of granitic grit.

Morris (1967) described the overlying El Gallo Formation 25 km west (loc. 20) as follows:

Near the coast the major beds containing dinosaur bones consist of brown to black lignitic shales and siltstones intercalated between massive, cross-bedded arkosic arenites. Conglomeratic lenses are common. The abundance of petrified wood is a distinctive characteristic of the arenites. Logs 3.6 to 5 meters long are not uncommon. Bones found in the arenites are generally disarticulated, and most show evidence of abrasion. In general, the more articulated hadrosaurian material is found in the shales and siltstones underlying the coarser arenites. Eastward but still in the Arroyo Del Rosario, the El Gallo Formation becomes predominantly coarse sandstone and conglomerate, and fossils are much rarer. The upper part of the formation near the unconformity separating it from the Rosario Formation contains a few isolated carbonate lenses rich in marine invertebrates. Occasionally, ammonite fragments are found in the lower shales and arenites. These were either reworked from older units or were washed in from nearby marine areas.

The geography during the time of deposition of the El Gallo Formation was very similar to that of the present, and the Cretaceous strand line was probably not far from the modern one. The lithology of the El Gallo Formation is indicative of near-shore lagoons and playas, a normal habitat for hadrosaurian dinosaurs. Vegetation was much more profuse than that which exists today, and the better-drained areas were thickly wooded. Streams entering the topographic lows deposited coarse fluviatile sediments over the finer materials of the lagoons and playas. The influence of fluviatile processes over marginal marine was more pronounced eastward toward the interior of Baja California where a region of higher relief persisted.

Farther southwest (loc. 21) McGee (App. 2 [1965]) analyzed the fauna in a sequence of alternately marine and nonmarine rocks of the Punta Baja Formation. He concluded that the marine environment was sublittoral (depth, 50 to 200 m) with rapid deposition.

Acosta (App. 1; 1970) studied 1,225 m of Upper Cretaceous sandstone and conglomerate around Bahía Soledad at the mouth of Valle Santo Tomás. His studies of sandstone fabric, mineral assemblages, and sedimentary structures led him to conclude (p. 30) that the sediment was derived from a nearby, tectonically active source of high topographic relief; this allowed streams to bring coarse material into a coastal area where it was slightly reworked and efficiently sorted. Here and directly south, the Upper Cretaceous strata were deposited against a steep and irregular coastline.

At the head of Valle El Morro (loc. 8) south of Mesa Redonda, the Late Cretaceous seas lapped into rock-walled canyons where rudisted clams clung to the angular blocks of bed rock at the foot of the sea cliffs. One kilometer west, the modern arroyo walls expose lenses of angular grit and boulders full of carbonized plant debris interbedded with claystone containing abundant baculites and lenses of shell debris. About 2.5 km west of the old sea cliff, McGee (App. 1) studied 105 m of marine strata that were deposited adjacent to a shallow submarine ridge. McGee wrote (p. 40),

Elements of the *Haplophragmoides* spp. and *Praebulimina joaquinensis* faunules suggest an outer shelf or, more probably, an upper bathyal environment with increase in water depth upwards through the stratigraphic section. The *Haplophragmoides* spp. faunule inhabited a restricted environment characterized by conditions conducive to the proliferation of arenaceous forms. Oxygen deficiency may have been important.

The environment was also detrimental to calcareous faunas, or to their preservation during burial.

Bailey (App. 1) studied the sedimentary structures of 670 m of the Rosario Formation along the coast north of Ensenada (loc. 9). He made the following conclusions (p. 56):

(1) The general paleocurrent-paleoslope direction is southwesterly varying considerably from place to place; (2) current directions of associated structures, formational members, and areas are related; (3) turbidity currents and slumping are the main agents of sedimentary structure formation; and (4) the environment of deposition indicates a near-shore, shelf-type environment.

Bailey's determination of a westerly or southwesterly paleoslope and transport from the east or northeast agrees with the findings of Worthington (App. 2) and Hord (App. 2) in the El Morro (loc. 2) and Arroyo Rosarito (loc. 1) areas and Maytum and Elliott (1970) in San Diego County.

The widest outcrop of Upper Cretaceous strata is near Mesa San Carlos (loc. 23). Here, the lower part of the section ranges from nonmarine to deep bathyal. The upper part is equivalent to and probably younger than Kilmer's type Rosario Formation. Northwest of Puerto San Carlos, the formation includes a unit that may be the result of submarine sliding. Blocks of Upper Cretaceous shale and sandstone, as much as 30 m in greatest dimension, are chaotically oriented in a matrix of rounded cobble conglomerate, shale, and sandstone. The unit itself is nearly flat lying, but the included blocks dip at all angles from flat to vertical and overturned. No megafossils have been found in either the blocks or the matrix. The structural confusion in this seaward exposure recalls similar descriptions by Anderson and Hanna (1935, p. 13) of Upper Cretaceous strata on Isla Navidad, 12 km south of Isla Cedros.

Vigorous erosion, rapid sedimentation, and, west of the Santillán y Barrera line (see below), rapid subsidence occurred in latest Cretaceous time. The shoreline fluctuated across a narrow continental terrace as rates of sedimentation and subsidence varied. Most of the sand is arkosic with many angular grains. Some is poorly sorted. Commonly, the source rock is found within a few kilometers of the site of deposition. This interval of uplift spanned 10 to 20 m.y. Flynn (1970) reported a high-relief unconformity between the Redonda and Rosario Formations, and McGee (App. 2 [1965]) illustrated a marked angular unconformity between the lower nonmarine and the lower marine units at Punta Baja.

Lindgren (1890, p. 29) recognized that the belt of Upper Cretaceous rocks marked a significant structural lineament. Beal (Marland Oil Company of Mexico, 1924) described the Upper Cretaceous rocks in terms of his "Baja California syncline," a giant downbuckle between the stable crystalline peninsula and the basement rock exposed in the Vizcaíno Peninsula, Isla Cedros, and the Channel Islands of southern California, U.S.A. Santillán and Barrera (1930, p. 7-8) clearly recognized the existence of a major tectonic hinge line:

De acuerdo con los párrafos anteriores, se puede concluir que dentro de la región estudiada aparecen rocas ígneas, sedimentarias y metamórficas, distribuidas en forma de zonas longitudinales que tienen contornos más o menos irregulares, pero que guardan la orientación general de la Península. Así, partiendo de la parte más elevada hacia la costa, o sea del este al oeste, se presenta, primero, la zona de las rocas graníticas, cuyo borde occidental puede limitarse por una línea que se alejara unos 10 km de la costa en la porción de Ensenada, y unos 40 km al este de Rosario, siguiendo una trayectoria más o menos recta, pero que se aleja de la costa a medida que se camina hacia el sur.

Este borde occidental de las mesas corresponde al antiguo litoral, el que tierne un contorno ondulado e irregular, permaneciendo a veces más o menos paralelo a la costa actual o bein teniendo una forma lenticular. En las partes donde este contorno se aleja de la costa, la planicie costera intermedia se halla cubierta por suelos arcillosos, como en el valle de San Quintín, o bein por areniscas no consolidadas, como se observa principalmente al sur de El Socorro.

Gastil and Allison (1966) pointed out that this remarkably straight line has controlled the depositional history of the Baja California coast for at least the past 100 m.y. We will hereafter refer to this as the "Santillán y Barrera line."

LOWER TERTIARY ROCKS

Stratigraphic Nomenclature

Marine Paleocene strata occur in many places along the coast from Punta San Isidro (loc. 11) west of San Vicente to the Vizcaíno desert (loc. 38) and again south from the western cape areas in the territory of Baja California (Beal, 1948; Mina, 1957). Marine strata of early Eocene age are found in several localities overlying the Paleocene strata. Marine middle or upper Eocene strata have been found only in the northwest corner of the state of Baja California (loc. 3), east of Rosarito Beach where there are no Paleocene rocks.

The marine Paleocene and Eocene strata are deltaic or nearshore deposits that in many places grade inland to brackish and freshwater deposits. There are also extensive deposits of nonmarine, postbatholithic, prevolcanic sedimentary rocks of unknown age.

The exposed Eocene-Paleocene deposits of the southern third of the state of Baja California fringe the coastline south from Punta Canoas (loc. 27), extending inland in the area of Bahía Santa Rosalía with possible correlative nonmarine deposits widely distributed throughout the interior of the peninsula. These nonmarine deposits are probably continuous beneath the Vizcaíno desert with the exposures of Tepetate Formation (Beal, 1948) in the southern half of the peninsula. North of Punta Canoas, the Paleocene rocks are exposed at elevations up to 600 m above sea level. They are found in the mesas along the coast and extend inland as much as 25 km near Mina San Fernando (loc. 24) and Mesa Purgatorio (loc. 19) north of El Rosario. North of Mesa Purgatorio, Paleocene rocks are found only in limited exposures near the coast. The relation of these strata to the unfossiliferous fluvial and lacustrine deposits that underlie the Miocene volcanic rocks in the eastern half of the peninsula is unknown.

In the northwestern part of the state, inland from Rosarito Beach (loc. 3), Eocene marine strata are found near the Pacific coast, and presumably Eocene fluvial deposits extend east to the crest of the Sierra Juárez (loc. 6). These resemble the Eocene rocks of the San Diego area (Kennedy and Moore, 1971) north of the international border.

In the southern part of the state of Baja California, Heim (1922, p. 534-535) named rocks that he thought were upper Eocene (now known to be Paleocene) the Tepetate Formation for Rancho Tepetate at lat 24° 23′ N.

Darton (1921), on the basis of invertebrate megafossils, considered that the rocks south of lat 30° N. were of Martinez age (Paleocene) and those north of lat 30° N. were of Tejon age (Eocene).

Santillán and Barrera (1930, p. 14-20) named the Paleocene (and Eocene) strata between Punta San Isidro and Mesa San Carlos Formación Sepultura for Mesa Sepultura (loc. 22) southeast of El Rosario. Beal (1948, p. 44-51) applied the name Tepetate to all Paleocene strata in the peninsula.

Mina (1956, 1957) used Formación Santo Domingo for Paleocene strata in the Pozo (well) Iray no. 1, 270 km north of La Paz; Formación Tepetate for Paleocene-Eocene strata in the region adjacent to the Tepetate type area, and Formación Malarrimo for strata along "la costa de Malarrimo" (northeastern Vizcaíno Peninsula). He named the Eocene strata of the Vizcaíno Peninsula Formación Bateque for Rancho Bateque, 32 km southwest of San Ignacio. Current work in the Vizcaíno Peninsula (John Robinson and John Minch, 1974, oral commun.) indicates that the Formación Malarrimo strata are Upper Cretaceous and the Formación Bateque includes Paleocene as well as Eocene strata.

Sedimentary Rocks

Fife (App. 1) described the southern coastal marine Paleocene and Eocene(?) strata of the Bahía Santa Rosalía quadrangle (p. 28):

In the northwest corner of the area near Punta María [loc. 31] up to 240 feet of marine Paleocene rests on granitic and metavolcanic rocks. The lower 100 feet of the section is mainly white to moderate reddish brown sandstone with minor conglomerate and limestone. It contains abundant echinoid spines, *Venericardia* sp., and *Turritella pachecoensis*. . . . The sandstone also contains many white siliceous concretions. They are usually spherical and about one inch in diameter.

This sandstone is identical to the presumably continental rocks to the east.

Farther east and higher in the section, the sandstone grades into cobble conglomerate composed of granitic and metavolcanic clasts. Gryphoid oysters were collected from the conglomerate. Again from Fife (p. 29-31):

The sandstone unit exposed at Punta María extends southward along the coast for several miles. At El Cordón, seven miles south of Punta María, the sandstone . . . exhibits a rich echinoid fauna.

Numerous small embayments of similar strata are found along the coast. South of Punta Negra they are usually capped by Pliocene strata.

The most extensive outcrop of fossiliferous Paleocene marine rocks occurs in the southern part of the area, to the north and in the vicinity of Rancho San Xavier. In this region at least one-hundred feet of dusky yellow to reddish brown sandstone is interbedded with conglomeritic sandstones, concretionary lenses, siltstone, and mudstone. Several horizons contain *T. pachecoensis* almost exclusively and form resistant strata. Locality B5D-30 yielded *Cerithidea* sp., *Ostrea* sp., *Ostrea* sp. nov., *Venericardia* sp., *Glycymeris* sp., and *Turritella pachecoensis*, plus several specimens of gryphoid oyster and unidentified horn corals, ostracodes, and foraminifera. This assemblage represents a sublittoral facies.

Farther north, about six miles up the arroyo from El Muertito . . . the fossil assemblage suggests a lagoonal or littoral environment. At this location the remains of a small proto-horse, *Hyracotherium* sp. nov., were interbedded with seeds from the family Chenopodieaceae (D. A. Preston, 1966, personal communication) *Cerithidea* sp., *Calyptraea* sp. nov., and *Ostrea* sp. . . .

About six miles due north of the above locality, at Occidental Buttes, Morris (1966) reported the discovery of ungulates of the orders Tillodontia, Perissodactyla and Pantodonta. Pantodonts of the family Barylambdidae were found stratigraphically above specimens assigned to Tillodontia and Perissodactyla. . . .

Morris (1966) identified *Esthonyx* sp. nov., of the order Tillodontia, and *Hyracotherium* sp. nov., of the order Perissodactyla.

Fife (p. 63-64) interpreted the Paleocene conditions:

The coast north of Punta Santa Rosalía [loc. 32] was rocky as it is today. The location of fossiliferous deposits indicates that heavy shelled gryphoid oysters were deposited with conglomerates near shore; while turritellas and echinoids were preserved with finer clastic sediment in the shallow embayments . . . .

South of Punta Santa Rosalía at least three major embayments existed. . . . The southern embayments are well exposed and contain shallow to brackish water faunas. These were areas of mud to sand bottoms. The positions of fossil mollusca, corals, ostracodes, and foraminifers suggest that alternate shallow marine and estuarine or brackish water conditions existed. Seeds of the family Chenopodieaceae and the remains of the proto-horse *Hyracotherium* are found in beds interfingering with typical estuarine strata. Ostracodes and charophytes indicate a lacustrine environment in the continental beds east of Rancho La Bachata.

These early Cenozoic rocks extend south from the Bahía Santa Rosalía quadrangle as far as Mesa Mesquital (loc. 38) where they disappear beneath the Desierto de Santa María (Llano del Berrendo). Inland, related sandstone lies beneath the Miocene volcanic rocks in the Sierra Santa Gertrudis (loc. 39), Sierra Sandia (loc. 37), Sierra San Borja (loc. 35), and north along the central valley of the peninsula to Cerrito Blanco (loc. 29). Isolated exposures of possibly related rocks crop out from the Sierra de la Asamblea (loc. 30) eastward to the gulf, on Isla Tiburón, and in the state of Sonora.

Typical and easily accessible exposures are those at Cerrito Blanco (loc. 29), a basalt-capped mesa about 3 km northeast of El Rosarito (loc. 34), the cliffs north of Misión San Borja (loc. 35), and the canyon above Misión Santa Gertrudis (loc. 39). At all four of these localities, the strata lie approximately horizontally. At the second and third localities, the sections are approximately 200 m thick. Typically, these are clean, quartz-rich, well-indurated sandstone beds, with siliceous cement. Their most prominent feature is cross-bedding, commonly in sets 2 m or more thick. Concretions occur in many places. They are characteristically the shell-in-shell type, 1 to 5 cm in diameter; in some places, they are so crowded that the strata consist entirely of various-sized spheres. Böse and Wittich (1913, p. 353) wrote:

En dos lugares más hemos podido observar que el terciario marino pasa al lado oriental de la Península, a saber: en el puerto de San Juan (San Borjas) y en el de Santa Isabel (cerca de Calmahí). [The Santa Isabel referred to here is presumably the one near Cerro Sandia.] En estos lugares coronan los depósitos terciarios la cumbre de la sierra grande granítica que forma el espinazo de la Península. Las rocas terciarias tienen aquí una potencia de unos 300 metros y quizá más.

We were unable to confirm the existence of marine strata in this area. Similar prevolcanic sandstone is found on the east side of Valle de San Felipe (loc. 16) and in the northern foothills of the Sierra Santa Rosa (loc. 15) west of San Felipe (Andersen, App. 1).

The age of the cross-bedded sandstone of the interior is unknown and remains a major question in the stratigraphic history of the area. Fife (1969) and Andersen (App. 2) attributed a Miocene age to these strata on the basis of the contained fragments of unmetamorphosed volcanic rock. There is evidence, however, for considering some of them to be of Eocene and Paleocene age:

1. Fife (App. 1) reported that unfossiliferous, white concretionary sandstone, identical with that found inland, occurs within the marine Paleocene section.

2. Eocene-Paleocene volcanism is known to have occurred in the peninsula (Gastil and Krummenacher, 1970).

3. North of lat 28° N., nonmarine deposits of demonstratably Miocene age are absent, as opposed to the known occurrences of nonmarine Paleocene and Eocene strata (Morris, 1966; Flynn, 1970; Minch, 1970).

4. On the basis of lithologic similarity to rocks of known age in the territory of Baja California, Beal (1948, p. 46) preferred a Paleocene age for the sandstone at Santa Gertrudis.

Paleocene and lower Eocene sedimentary rocks are well exposed along the Pacific coast of the central part of the state. Unlike in the southern part, where lower Tertiary deposits lie directly on basement rocks at low elevation and reflect a topography and coastal geography closely akin to that observed today, here they rest on a surface of Upper Cretaceous strata that is as much as 350 m above sea level. The surface rises toward the interior of the peninsula where Paleocene and Eocene strata abut basement rocks.

Santillán and Barrera (1930, p. 14, 19) included all the Paleocene and Eocene strata in their Formación Sepultura, with the type location Mesa Sepultura (loc. 22) east of El Rosario:

FORMACION SEPULTURA. (*Eocene inferior*).—Descansando directamente sobre la formación Rosario y en concordancia con ella, se observa en varios lugares una serie de rocas sedimentarias, que hemos llamado "Formación Sepultura," porque en la porción de terreno donde se encuentra este cerro, es donde esa formación está mejor caracterizada. Las rocas que la constituyen son principalmente areniscas de colores gris, verde, rojo y amarillo, alternando con capas de conglomerado de tamaño medio uniforme y también con margas calizas. Algunas veces aparecen, intercaladas en las capas de arenisca, capas de arcillas compactas amarillentas.

Las areniscas de esta serie son muy parecidas a las de la formación Rosario, pues son generalmente graníticas y los elementos constitutivos esenciales son cuarzo, feldespatos y minerales ferromagnesianos (principalmente mica y epidota) comúnmente alterados. Hay lugares donde la semejanza de las areniscas entre las formaciones Rosario y Sepultura, así como el paralelismo de los planos de estratificación es tan grande, que no es fácil distinguir los límites de cada serie. En la arenisca gris se encuentran con frecuencia nódulos lenticulares de arenisca muy compacta, algunas veces con fósiles y también con pedacitos de materia carbonosa. Este tipo de arenisca pudimos observalo en la costa al norte de Punta Colnett; pero no aparece en el cerro de la Sepultura, situado a unos 30 km al E. de Rosario. . . .

Los conglomerados no son tan compactos como los de la formación Rosario pero sí llegan a adquirir espesores considerables, como acontece en Punta Colnett. Las rocas principales de que están constituídas son: metamórficas, andesíticas, graníticas en menor proporción, cementadas por areniscas de grano grueso. Dentro de estos conglomerados es frecuente también observar el color rojizo y en ellos encontramos algunos fósiles.

Las capas de marga caliza no son abundantes ni de gran espesor; solamente pudimos observarlas en el cerro de la Sepultura, con un espesor aproximado de unos 2 metros. Esta capa se presenta bastante compacta y también contiene fósiles, aunque no fácilmente determinables por estar muy quebrados.

Santillán and Barrera (1930, p. 20) reported the following fossils: *Turritella pachecoensis* Stanton; *Retipirula crassitesta* Gabb; *Turritella martinezensis* Gabb; *Hercoglosa* n. sp.; *Brachiospingus* sp.; *Nática* cf. *pinonensis* Dickerson; *Turritella infragranulata* Gabb; *Polinices hornii* Gabb; *Amauropsis martinezensis* Gabb; *Macrocallista* cf. *stantoni; Glycimeris* sp.; *Venericardia venturaensis; Cucullea mathewsonii* Gabb; *Tellina undilifera* Gabb.

Unlike the marine Paleocene rocks in the southern area, this formation is usually conglomeratic. In its most westerly exposures on Mesa San Carlos, however, conglomerate forms only channels in mudstone.

Beal (1948, p. 45) reported 670 m of Paleocene strata west of Mesa San Carlos. We have measured sections at several points adjacent to Puerto San Carlos; if this is the area described by Beal, he did not distinguish between Upper Cretaceous and Paleocene strata. We did not find more than 300 m of Paleocene rock in this area.

Inland from Mesa San Carlos, cliffs expose the Cenozoic section for almost 30 km. The Paleocene section becomes progressively thinner, more coarsely clastic, and less fossiliferous to the east. The most easterly fossils found are diminutive, thin-shelled oysters. Farther inland, stratigraphically equivalent unfossiliferous, presumably nonmarine, strata wedge out against the irregular buried topography.

Eastward at a higher elevation (loc. 24), however, we found more than 40 m of pale, bluish-gray siltstone (Fig. 22) with an early Eocene foraminiferal fauna

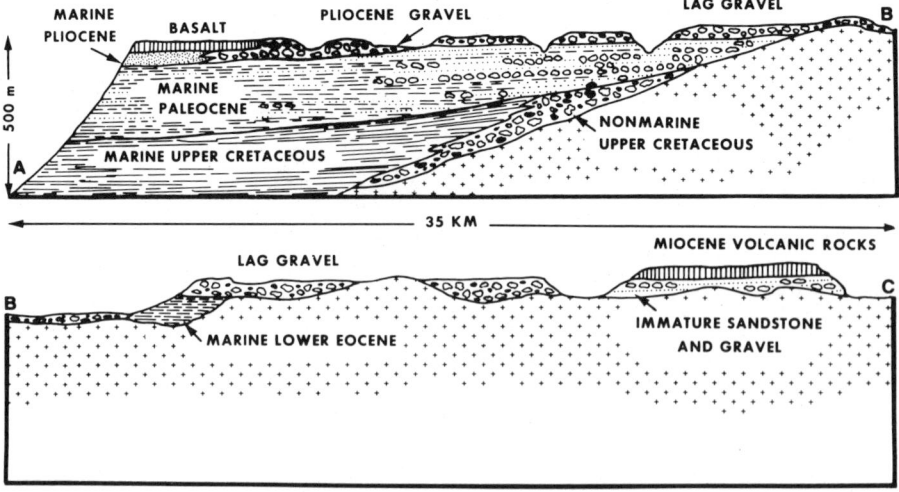

Figure 22. Diagrammatic section ABC through Mesa San Carlos from the Pacific coast to east of El Mármol (locs. 23 through 26 in Fig. 20).

indicating a depositional depth of at least 160 m. These Eocene strata are in turn overlapped by poorly bedded conglomerate, interpreted as lag gravel, that appears to drape unconformably on all of the early Tertiary and older rocks. The source of the lag gravel may have been Eocene or younger strata that have been eroded away.

The position of the lower Eocene marine strata raises questions concerning the paleogeography. If 160 m of water were above these deposits today, a seaway would stretch across the peninsula. It is possible that the sea once extended farther inland and into the area now occupied by the Gulf of California. Even if the Eocene sea did not extend farther inland, the erosion surface that lies beneath the immature fluvial and lacustrine strata in the eastern half of the peninsula must have been near sea level during Eocene time.

Northward 80 km at Mesa Purgatorio (loc. 19), Kilmer (1963) mapped a large area of Paleocene rocks capping the Upper Cretaceous strata at an elevation of about 627 m.

The most northern exposure of Paleocene rocks known in Baja California is at Punta San Isidro (loc. 11; Santillán and Barrera, 1930). The rocks represent a shoreline environment, at the base of a steep sea cliff, similar to that found in the area today. Paleocene strata have not been recognized between Punta San Isidro and the Santa Ana Mountains, southern California.

In the interior of the peninsula, north of Arroyo San José, Miocene volcanic strata generally overlie immature sandstone and conglomerate of undetermined age. These strata rest unconformably on an irregular surface of Mesozoic basement rocks that in many places project through the sedimentary and volcanic rocks. They may be continuous with the southern (Paleocene-Eocene[?]) sandstone and with the isolated exposures of prevolcanic cross-bedded sandstone near San Felipe and Valle San Pedro. Apparently, when Miocene volcanic rocks first erupted, the peninsula east to the Gulf of California was extensively covered by nonmarine sandstone and conglomerate with isolated hills and ridges of bed rock separated by broad valleys as in northwestern Sonora today. The erosion surface that was formed in Late Cretaceous and possibly Paleocene time was mantled with continental deposits during Eocene time when the interior valleys of the peninsula stood little above sea level.

Gabb (1882) referred to the Miocene volcanic strata as he did the Paleocene rocks near Bahía Santa Rosalía as "post-Pliocene." Emmons and Merrill (1894, p. 508) noted these strata in the region of El Mármol (loc. 26):

Here the divide line is marked by occasional isolated table-topped buttes, capped by rhyolite, which rise 500 to 1,000 feet above the desert level and serve to mark the original level of the mesa sandstones, which have been protected from erosion by the cap of more enduring rock. These rhyolites are generally of earlier date than the lake beds. The top of the mesa sandstones as thus determined is about 3,000 feet above present sea level, and their maximum observed thickness 800 feet.

Our thickest measured section was 210 m near Santa Ynez.

Emmons and Merrill used Gabb's name Mesa Sandstone for the flat-lying sandstone and conglomerate underneath the Miocene volcanic rocks. This was reasonable in the light of Gabb's use of the term in its type area in the southern part of the peninsula. Gabb (1882, p. 147), however, used Mesa Sandstone in the El Mármol area to indicate folded and metamorphosed basement rocks and used "post-Pliocene" for the flat-lying unmetamorphosed strata resting on them.

Middle and upper Eocene rocks are extensively exposed in San Diego County, California. M. A. Hanna (1926) recognized the La Jolla and Poway Formations of the La Jolla quadrangle north of San Diego to be of middle and late Eocene age, respectively. Kennedy and Moore (1971) and Peterson (1970a) discussed the stratigraphy of these rocks.

Rocks of proven middle to late Eocene age are scarce throughout the state of Baja California. Beal (1948, p. 50) reported upper Eocene (Tejon) strata at one locality 8 km southeast of Punta Colnett (loc. 14) and at another locality 8 km inland from San Quintín (loc. 17). We were unable to relocate either. The strata are presumably included in Paleocene and Upper Cretaceous rocks as shown on Plate 1. Perrilliat-Montoya (1968) described a fossil locality 8 km southeast of San Quintín that she attributed to the Tepetate Formation.

Well-exposed middle and upper Eocene strata were described by Flynn (1970) from localities inland from Rosarito Beach. Two new formation names, Delicias Formation and Buenos Aires Formation, were defined (Flynn, 1970, p. 1793-1797):

Delicias Formation. . . . This formation can be divided into an upper sandstone member represented in the type section of the formation and a lower mudstone member. . . .

The type section for the lower mudstone member is 4 km southeast of Rancho Delicias on the grade below Rancho Buenos Aires. . . .

The type locality of the upper sandstone member is in the amphitheater just south of Rancho Delicias. . . . Fossil beds are common in these sandstones. Characteristic species include *Ostrea* sp., *Potamides carbonicola* (Cooper), *Amaurella* sp., *Scolimytilus* sp., and *Pelecyora* sp. (S.D.S.C. [San Diego State University] fossil locality no. 181). This fauna and its associated lithology indicate a restricted, shallow, brackish water depositional environment. The mudstones and sandstones are also lithologically and paleontologically similar to those of the Del Mar sandstone member of the La Jolla Formation in San Diego County, Alta California ([M.A.] Hanna, 1926).

Buenos Aires Formation. . . [This] series of densely packed cobble conglomerates and overlying sandstone beds [is] in the southeastern and central portion of the mapped area [southeast of Tijuana]. This formation may be divided into a lower conglomeritic member and an overlying sandstone member. The type section is located on the grade leading from Valle Cuero de Venado toward Rancho Buenos Aires. . . .

In the type section, the contact between the conglomerate and the mudstone member of the Delicias Formation is gradational. To the northwest, the Buenos Aires Formation rests unconformably on the basement complex and the Delicias Formation. . . .

[The] faunule appears to be characteristic of a sublittoral, tropical to subtropical environment of deposition. Gastropods comparable to *Turritella, Ectinochillus,* and *Ficopsis* of this fossil fauna are characteristic of the tropics or near-tropics. The presence of moderately thick-shelled pelecypods and the semi-parallel arrangement of the turritelloid forms in distinct layers indicates a moderate amount of wave action or current, or both. However, this environment was static enough to allow the deposition of abundant mica and the development of delicate thin-shelled gastropod associations.

The fossil fauna indicates an age for the Buenos Aires sandstone member which is late Eocene in the West Coast chronology. Correlations are with the upper part of the Rose Canyon Shale in San Diego County ([M.A.] Hanna, 1926), with the Tejon Formation in the southern San Joaquin Valley, with the upper portion of the Matilija and the lower Cozy Dell Formations of the Santa Ynez Mountains and Central Ventura Basin, with the Coaledo

Formation of southwestern Oregon, and with the lower portion of the Cowlitz Formation in Washington.

Sauer (1929), Flynn (1970), Minch (1970), and Frazer (App. 2) described conglomerate that resembles the apparently nonmarine conglomerate of the Buenos Aires Formation and extends eastward to the top of the Sierra Juárez. These rocks are well exposed along the Ensenada-Tecate highway between Vallecitos and Las Palmas (loc. 4), in the hills east of Rancho El Campadre (loc. 5), and in the area southwest of the town of La Rumorosa on Mexican Highway 2 (loc. 6).

These gravel deposits superficially resemble the Poway and Ballena gravel deposits in San Diego County. Both north and south of the border, the gravel deposits are marine to the west and fluvial to the east; they consist of rounded, resistant boulders and cobbles that produce ridges of reverse topography. Minch (1970), however, noted that the clast population of the southern "Las Palmas-type" gravel is distinctly different from the northern "Poway type." The Poway-like Ballena Gravel (Fairbanks, 1893; Miller, 1935b) lies at elevations of up to 1,000 m in the interior of the peninsula; the Buenos Aires-like conglomerate is found at elevations of up to 1,600 m and as far inland as the eastern escarpment of the Peninsular Ranges. The similarity rests in the geographic distribution of marine and fluvial facies and in the physical appearance of both the included clasts and the ridges of reverse topography. The clast population of the Las Palmas type (Minch, 1970) is distinctly different from the Poway type (Bellemin and Merriam, 1958; DeLisle and others, 1965; Woodford and others, 1968). In contrast to the typical Poway type, which is composed of perhaps 90 percent weakly metamorphosed rhyolite-dacite volcanic rocks and less than 10 percent metasedimentary rocks (primarily quartzite), the Las Palmas type (Table 8) is rich in weakly metamorphosed quartzite, arkose, wacke, and pebble breccia. There are quartz-bearing metavolcanic rocks, but only a small proportion of them are kinds that appear commonly in deposits of the Poway type. The eastern exposures of the conglomerate units of the Sierra Juárez, especially those near La Rumorosa, contain a considerable proportion of less resistant volcanic, sedimentary, and granitic rocks. Different but possibly analogous conglomerate units occur near Jacumba, San Diego County (Miller, 1935a; Brooks and Roberts, 1954); in the southern Sierra Juárez (Minch, 1970); and beneath the volcanic section in the Sierra Santa Rosa and the Sierra San Felipe (Andersen, App. 1).

Although these early Tertiary conglomerate units are widespread, most of them are found in well-defined channels that have been interpreted (Minch, 1970) as the beds of paleorivers that flowed across the peninsula from east to west (Fig. 23).

Ironically, the conglomerate units that lie on the high plateaus are best known in the literature for the often quoted observation of Lindgren (1888) at Campo Nacional (loc. 7), a small isolated hill southeast of Laguna Hanson, far removed from the major areas of conglomerate. Lindgren wrote as follows (p. 192-193):

Standing on a small elevation nothing but granite can be seen as far as the eye reaches, in all directions. Within a radius of a few miles the gulches leading down to the desert have been worked and yielded a considerable quantity of gold. The latter is coarse and well worn, and the gulches are filled with well rounded smooth pebbles of white quartz, or a dark quartzite.

This, in itself, is remarkable, as there certainly are no metamorphic rocks anywhere in the vicinity. It was soon found that all these gulches led up to a small flat-topped hill about 200 feet above the plateau, called the Black Hill. This hill is about one-half mile long, east and west, and one-eighth to one-quarter mile wide; it is made up of a well-packed mass of auriferous metamorphic gravel in very smooth boulders, often six inches in diameter. . . .

That this patch of auriferous gravel has been formed by an ancient river of considerable importance is certain, but the most interesting questions are, whence did it come and to where did it flow, and where are the metamorphic rocks that furnished the material for the boulders?

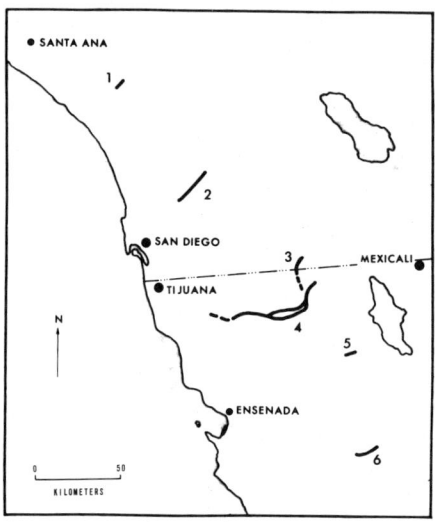

Figure 23. Location of channels of early Tertiary streams that flowed southwesterly in the northern Peninsular Ranges: 1, Santa Rosa River; 2, Ballenas River; 3, Jacumba River; 4, Rio de las Palmas; 5, Rio Nacional; 6, Rio del Rodeo (from Minch, 1970).

TABLE 8. PERCENT CLAST COMPOSITION OF EARLY TERTIARY GRAVEL UNITS BY AREA

TYPE	Santa Rosa *	Ballena (Poway type)	Jacumba	La Rumorosa–Las Palmas	Campo Nacional	El Rodeo
Quartz	x	0.5	3.7	2.5	13.5	1.5
Quartzite	x	6.7	12.2	21.5	6.0	..
Red Sandstone		0.2	..	3.0	..	..
Feldspathic Quartzite		1.2	1.0	2.7	..	8.5
Black Quartzite		1.2	3.1	3.5	42.0	9.5
Sandstone		..	0.7	1.2	..	..
Chert Pebble Conglomerate		..	..	3.3	..	..
Reddish Metarhyolite	x	67.3	..	6.9	..	..
Granitic Rocks	x	2.4	14.3	2.7	17.5	16.5
Argillite		0.8	0.8	1.2	..	1.5
Flow-banded red volcanic rock		..	0.1	1.1	..	..
Black Hornfels		..	0.6	0.6	..	..
Purple Metavolcanic rock		0.5	1.1	0.9	..	..
Aphanitic Metavolcanic granulite		7.5	21.4	19.5	..	..
Green Meta-Sandstone or Breccia		3.0	8.8	9.0	..	..
Green Metavolcanic porphyry		7.0	23.9	18.0	..	..
Brown Metarhyolite		..	..	0.6	..	..
Gray Metarhyolite		..	2.4	0.1	..	..
Gneiss		..	4.2	0.6	..	9.5
Chlorite Schist		..	..	..	..	16.0
Black Slate		..	..	..	11.5	10.0
Black Gneissic Quartzite		..	..	..	8.5	..
Black Chert		..	..	..	..	22.5
Miscellaneous		2.5	2.1	2.4	1.5	4.5

* No detailed analysis. "x" indicates present.
". ." indicates no clast of this composition found.

NOTE: from Minch, 1970.

Volcanic Rocks

Daniel Krummenacher and a group of students at San Diego State University determined in 1969 the K-Ar ages of 30 postbatholithic volcanic rocks from the state of Baja California (Table 9). An alkalic rock (sample KA-524, olivine, pyroxene, biotite, potassium feldspar) from the base of the Tertiary section south of Bahía de los Angeles (loc. 36) was dated at 59.0 ± 1.8 m.y. B.P. using the feldspar fraction. An andesite clast in the conglomerate underlying the Miocene volcanic rocks in the southern Sierra Juárez (loc. 12) was dated at 53.8 ± 1.5 m.y. B.P. using the biotite fraction.

Beal (1948, p. 111) and Mina (1957, p. 132) reported volcanic rocks of probable Eocene age in the southern half of the peninsula.

TABLE 9. K-Ar "AGES" FOR VOLCANIC ROCKS IN THE STATE OF BAJA CALIFORNIA

Locality[*]	Sample[†]	K-Ar Age[◊] m.y.	Mineral	Rock and Locality
31	524	59.0 ± 1.8	Oligoclase?	Basic alkaline flow, base of stratigraphic section southwest of Bahía de las Animas
12	533	57.8 ± 1.5	Hornblende	Cognate of pure green hornblende in an oxyhornblend andesite (B6G-11), Sierra San Felipe
9	563	53.8 ± 1.5	Biotite	Clast of andesite in basal Tertiary conglomerate, east of El Rodeo, southern Sierra Juárez
35	518	19.8 ± 0.5	Biotite	Rhyolite tuff, basal volcanic unit, Misión Santa Gertrudis
31	523	17.0 ± 0.5	Hornblende	Andesite boulder from conglomerate overlying rock 524 (above)
31	525	12.7 ± 0.5	Biotite	Tuff overlying 523
31	526	11.9 ± 0.5	Whole rock	Basalt overlying tuff 525
31	527	10.5 ± 0.4	Whole rock	Top of section, basalt overlying 526
31	522	10.0 ± 0.5	Whole rock	Basalt, southwest of Bahía de las Animas, stratigraphic relation to above sequence unestablished
31	531	5.1 ± 0.4	Plagioclase	Basaltic andesite from the Llano San Pedro, south of Bahía de las Animas
28	554B	14.2 ± 0.5	Biotite	Rhyolite tuff northeast of Misión San Borja
28	546	14.0 ± 0.2	Biotite	Same rhyolite tuff unit, northeast of Misión San Borja
28	554P	12.6 ± 3.0	Plagioclase	Same sample as 554B
28	553	12.0 ± 0.4	Whole rock	Basalt overlying 546
23	535	16.3 ± 0.5	Hornblende	Andesite resting on bed rock just south of El Mármol
7	539	16.1 ± 2.1	Plagioclase	Basaltic andesite flow, top of old grade south of Valle de la Misión, Tijuana quadrangle
9	549	13.6 ± 0.4	Whole rock	Basalt overlying 563 (above), southern Sierra Juárez
9	550	10.0 ± 0.5	Biotite	Welded rhyolite capping most of the southern Sierra Juárez, overlying 563 and 549
19	555	7.3 ± 1.5	Plagioclase	555 through 559 is a sequence up a section of steeply tilted acidic volcanic rocks northwest of Puertecitos
19	556	8.0 ± 1.0	Plagioclase	
19	557	9.4 ± 0.7	Plagioclase	(The overlap assigned by the plus and minus suggests a best guess of 8.7 - 8.8 m.y.)
19	559	8.3 ± 0.8	Plagioclase	
19	560	5.9 ± 0.2	Plagioclase (impure)	561 overlies 560 in a sequence of flat-laying siliceous pyroclastic units unconformably above 555 through 559
19	561	3.1 ± 0.5	Plagioclase	
11A	541	8.9 ± 0.6	Hornblende	Andesite, northern Sierra Pinta (collected by McEldowney, see App. 1)
11A	540	9.5 ± 1.0	Plagioclase	Dacite, unconformably above 541 (collected by McEldowney, see App. 1)
11	517	7.6 ± 0.4	Whole rock	"Basalt" from base of volcanic sequence near the southwestern edge of the Sierra Pinta (collected by James, see App. 1)
27	538	4.3 ± 2.0	Whole rock	Basalt, near coast, south of Arroyo San José, Punta Canoas quadrangle
32	552	2.6 ± 0.5	Whole rock	Youthful basaltic flow, east of Rosarito, Bahía Santa Rosalía quadrangle
3	800	18.5 ± 1.1	Plagioclase	800 - 801 are from the same andesite lahar breccia at Jacumba, California. This rock immediately underlies the basalt dated by Hawkins (1970a) at 18.7 ± 1.3 m.y.
	801	18.6 ± 0.8	Hornblende	

* See Figure 24 and Plate 1-A, 1-B, and 1-C.
† See Table 10 for chemical analyses and see Appendix 7 for petrographic description and locality description.
◊ Determined by Daniel Krummenacher, K-Ar Laboratory, San Diego State University.

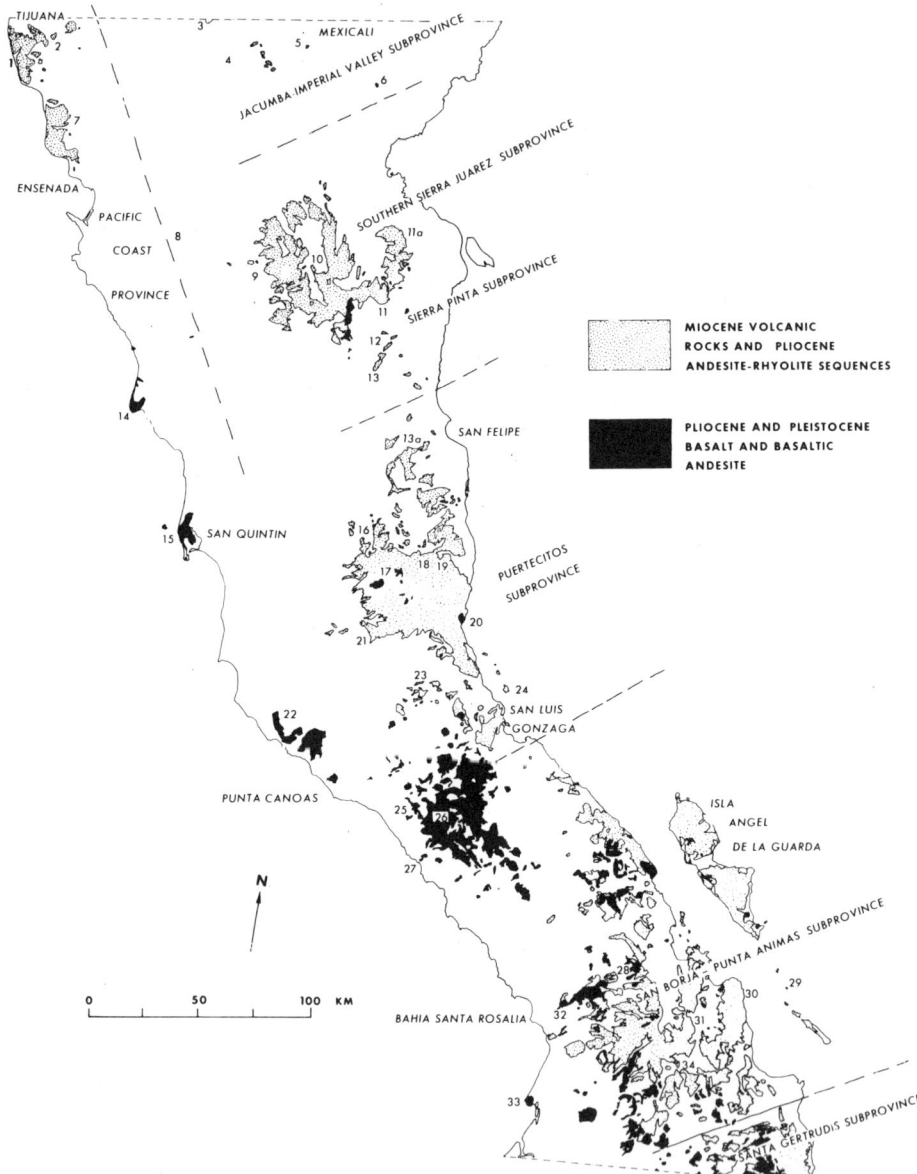

Figure 24. Distribution of Cenozoic volcanic rocks. Numbers 1 through 35 correspond to localities cited in the text.

5

Mid-Cenozoic Volcanism

During late Oligocene or early Miocene time, a new tectonic regime began. This was not a continuation or rejuvenation of the Mesozoic mountain building that encircled the Pacific Ocean but rather a dilational event extending from the Pacific Ocean to the Colorado Plateau. Block after block of the once stable granitic crust broke into elongate splinters that sank, tilted, and in some places squeezed against one another. Over this terrain of broken and tilted crust poured rhyolite, andesite, dacite, and basalt. Plutonic intrusion and metamorphism occurred locally. Today, the area affected by this event is called the Basin and Range province (Fig. 1). The Gulf of California and the eastern third of the state of Baja California are included in this province. During the Miocene Epoch, the gulf extended into areas where there had been no sea for more than 100 m.y. Ultimately, gulf waters lapped against the foothills of the Transverse Ranges and the eastern valleys of the Mojave Desert in California, U.S.A., and extended into Arizona and Sonora. Although late Pliocene and Pleistocene strata lie flat in many areas, suggesting recent stability, the gulf province is seismically and thermally active, implying that plutonic and metamorphic activity continues at depth.

STRATIGRAPHIC NOMENCLATURE

In the state of Baja California, Miocene marine strata have been identified in only two areas: the middle Miocene Rosarito Beach Formation (Minch, 1967; Minch and others, 1970) in the northwestern corner of the state (locs. 1 and 7 in Fig. 24, where all locations mentioned in this chapter are shown) and late Miocene rocks (Andersen, 1969) west of San Felipe (loc. 13a). In the first locality, the sedimentary strata are associated with basalt of the Continental Borderland province; in the second area, they overlie the volcanic rocks of the gulf depression. Nowhere in the state can the mid-Cenozoic volcanic suite of the Gulf of California province be related to the sedimentary record. In the southern half of the peninsula, however, the marine Miocene sedimentary rocks are extensive and laterally equivalent to great thicknesses of andesite.

Early writers, principally Heim (1915, 1921, 1922), Darton (1921), and Beal (Marland Oil Company of Mexico, 1924), used a variety of formation names to designate the marine Miocene strata in the south-central part of the peninsula (Fig. 25). The early work is summarized by Hertlein and Jordan (1927) and Beal (1948). Beal called the older strata the Purísima Nueva Formation (previously known as "Monterey Formation") and the San Gregorio Formation. These are overlain by sandstone, the Ysidro Formation, which was named by Heim in 1922 for the village of San Isidro (or San Ysidro) east of La Purísima. The Ysidro Formation is well exposed on both sides of the peninsula. It is found as far north as San Ignacio on the Pacific slope. It occurs on the gulf coast as far north as lat 25°30' N. (Beal, 1948, p. 68) and possibly as far as 60 km north of Santa Rosalía (Beal, 1948, p. 57).

Beal (1948, p. 57) stated that the Ysidro Formation "apparently becomes more tuffaceous toward the gulf coast." From Beal's discussion of the Ysidro-Comondú contact (p. 69-73) it is clear that the formations are in some areas separated by local erosional and even angular unconformities, whereas in other areas there is a continuous transition from the marine Ysidro sandstone units upward into progressively more volcanic detritus and true volcanic strata. Beal (p. 64) correlated the Ysidro Formation with the Vaqueros or Temblor Formations of California, U.S.A.

Heim (1922) introduced the name Comondú Formation for the volcanic sandstone, conglomerate, breccia, and included lava (basalt) overlying the Ysidro Formation near the village of Comondú (lat 26°04' N., long 111°50' W.). He found a maximum thickness of more than 1,330 m in the Sierra de la Giganta. Beal (1948) extended

	NORTH OF THE BORDER		STATE OF BAJA CALIFORNIA		TERRITORY OF BAJA CALIFORNIA	
	SAN DIEGO	IMPERIAL VALLEY	WEST COAST	GULF DEPRESSION	WEST COAST	GULF COAST
PLEISTOCENE	BAY POINT FM. 1	BORREGO FM. 7 6				
	LINDA VISTA FM. 2 = SWEITZER FM. 1	PALM SPRING FM.	LINDA VISTA FM. 10	UNNAMED MARINE STRATA AND VOLCANIC ROCKS 12		
PLIOCENE	SAN DIEGO FM. 3 4	6	SAN DIEGO FM. 10		SALADA FM.= ALMEJAS FM. 13 14	INFIERNO FM. = 16 MARQUER FM. 8
		IMPERIAL FM. 8	CANTIL COSTERO FM. 11			GLORIA FM.= 16 CARMEN FM. 9
		CANEBREAK CONGLOMERATE				BOLEO FM.= 16 SAN MARCOS FM. 8
				UNNAMED MARINE 12	ATAJO FM.= COMONDU FM.= 13 SAN ZACARIAS FM. 14	
		SPLIT MOUNTAIN FM. 9		UNNAMED VOLCANIC AND NONMARINE STRATA 12	TORTUGA FM. SANTA CLARA FM. 14	HORNILLAS FM. 17
MIOCENE	OTAY FM. 5		ROSARITO BEACH FM. 10		SAN IGNACIO FM. 14	
		ALVERSON ANDESITE 9			ZORRA FM. 14 YSIDRO FM. SAN JOAQUIN FM. 14 15	PILARES FM. 17
		ANZA FM. 9				MINITAS FM. 17
						PELONES FM. 17
OLIGOCENE				NONMARINE DEPOSITS PROBABLY OF THIS AGE	SAN GREGORIO FM.= 15 PURISIMA NUEVA FM. 13	SALTO FM. 17

Figure 25. Correlation of middle to upper Cenozoic formations. Numbers opposite formation names indicate the following sources: 1, Hertlein and Grant (1939); 2, M. A. Hanna (1926); 3, Dall (1898); 4, Ingle (1973); 5, Artim and Pinckney (1973); 6, Dibblee (1954); 7, Tarbet (1951); 8, Durham (1950); 9, Woodard (1974); 10, Minch (1967); 11, Beal (1948); 12, Andersen (App. 1); 13, Heim (1922); 14, Mina (1957); 15, Beal (Marland Oil Company of Mexico, 1924); 16, Wilson and Rocha Moreno (1955); 17, McFall (1968).

the use of the term the length of the peninsula and used it to include all volcanic and associated rocks that overlie the Ysidro Formation and underlie the Salada or equivalent formations. Beal wrote (p. 74-75):

The Comondú formation is composed principally of volcanic rocks and fragmental rocks derived from them and is the record of a period, the time span of which has been determined only by inference, of widespread and intense volcanic activity, followed in many places by very active erosion. . . . It includes large porphyritic and dense intrusives, rhyolitic, andesitic, and basaltic lavas and tuffs, agglomerates, and mud flows as well as conglomerates, sandstones, clays, and some limestones. . . .

The formation varies greatly in thickness. In some places it is represented by a single lava flow less than 50 feet thick, yet it forms the greater part of the mass of the sierra in the central part of the peninsula and in this region attains a thickness of over 4000 feet. It occupies about one-sixth of the peninsula.

North of 28° N. Lat. the areas occupied by rocks assigned to the Comondú formation are much smaller and of minor importance as compared with those farther south.

McFall (1968) used the term Comondú Group for a sequence of late Oligocene to Miocene volcanic and volcaniclastic strata more than 4,000 m thick in the Bahía Concepción area of the gulf coast (lat 26°30' to 26°57' N.). He designated the basal unit as follows (p. 7):

Salto Formation. Somewhat over 1,000 feet of red, cross-bedded, tuffaceous sandstone with interbedded light-colored tuffs unconformably overlie the basement complex. It is here proposed to call these beds the Salto Formation and to designate as type section its exposures along the mountain front just north of the mouth of Arroyo Amolares, approximately 2 miles north of Rancho Salto in the west-central part of Concepcion Peninsula. The contact of this formation with the overlying Pelones Formation is arbitrarily designated as the base of the lowest significant bed of agglomerate.

McFall divided the Comondú Group into the basal Salto Formation (300 m of tuffaceous sand), the Pelones Formation (1,800 m of flows and ejecta), the Minitas Formation (150 m of coarse conglomerate), the Pilares Formation (90 m of basalt flows), the Hornillas Formation (150 m of conglomerate), and the Ricasón Formation (1,500 m of agglomerate, lava, and tuff). The Salto Formation is dated by a biotite tuff at 28.1 ± 0.9 m.y. B.P. The Pelones, Minitas, and Pilares Formations predate tonalite that is 20 ± 2.0 m.y. old; the Ricasón Formation postdates the tonalite.

McFall's Salto and Pelones Formations (bracketed between 28 and 20 m.y. B.P.) are older than the volcanic strata of the type Comondú. They must be the temporal equivalents of the tuffaceous Ysidro Formation and even the underlying Gregorio Formation. The Ricasón Formation may well correlate with the more extensive upper Miocene volcanic strata found elsewhere in the peninsula.

In the state of Baja California, the sedimentary strata underlying and interbedded with the mid-Cenozoic volcanic rocks are nonfossiliferous. The widespread sandstone that underlies the volcanic section may be Paleocene or Eocene in age (see Chap. 4).

Southwest of Bahía de las Animas (loc. 31), the alkalic volcanic rocks dated by the K-Ar method at 59 m.y. B.P. are overlain by red sandstone resembling that described by Beal (1948, p. 73), which is in turn overlain by conglomerate composed primarily of volcanic clasts. An andesite clast from the conglomerate was dated at 17 m.y. B.P.; from that and the fact that the conglomerate is overlain by volcanic rocks 13 m.y. old, it must be of Miocene age.

It is possible that marine Miocene strata exist in the interior of the northern peninsula. Two reports in the literature indicate that oysters have been found in sedimentary strata somewhere near Misión San Fernando: one by Böse and Wittich (1913, p. 352) and the other by Flores and González (1913, p. 246), possibly describing the same locality (App. 6). The stratigraphic position and poor preservation described suggests that they were not recording midden deposits.

Although Beal extended the name Comondú to the volcanic rocks in the state of Baja California, preliminary K-Ar dating (Table 9) suggests that the rocks of the Bahía Concepción-La Giganta-Comondú region in the territory of Baja California are of middle Miocene to Oligocene age, whereas those of the state of Baja California are of middle Miocene to Pliocene age. Reconnaissance petrographic and chemical comparisons indicate that the mid-Cenozoic southern sequence is composed predominantly of andesite and basalt, whereas the comparable northern sequences are andesite to rhyolite.

VOLCANIC ROCKS

Previous Observations

Gabb, Lindgren, Woodford, and Beal largely avoided the great volcanic accumulations north of lat 28° N. Beal's map shows the state almost devoid of Cenozoic volcanic rocks. Yet volcanic rocks were mentioned by Emmons and Merrill (1894, p. 508), Johnson (1924), and Flores and González (1913).

Hirschi and de Quervain (1927-1933) collected volcanic rocks from the Sierra Pinta and the coastal areas from Bahía Gonzaga to El Barril. Böse and Wittich (1913, p. 356) described the volcanic rocks in the southern mountains of the state and made the following significant observation (p. 354): "En las costas del pacífico hemos encontrado únicamente basaltos, en el lado del Golfo de California también andesitas."

Gabb (1882, p. 141), Heim (1922, p. 543, 547), Darton (1921, p. 746), and Anderson (1950, p. 47-48) cited evidence that led them to conclude that the source of the Comondú volcanic rocks was along the eastern edge of the peninsula if not indeed beneath the area now occupied by the Gulf of California. Beal (1948, p. 76), however, believed that "the locus of volcanism may have been in the gulf not far from the coast, but there [was apparently] no good reason why it could not have centered at the latitude of La Giganta, near the axis of the present uplift."

All of the writers from Heim (1922) to Mina (1956, 1957) regarded the Sierra de la Giganta on the crest of the peninsula as the greatest thickness of volcanic strata (1,500 m according to Mina). The mapping of McFall (1968) in the territory of Baja California, however, showed that the section along the gulf was at least 4,000 m thick. Our reconnaissance work in the state of Baja California has also shown that the volcanic section is thickest in the Gulf of California depression.

Terminology

Our field terminology for volcanic rocks was based on visible phenocrysts, color, fracture, rock structure, and mode of occurrence. The names so derived generally determine the map-unit descriptions but can differ from the rock name based on chemical analysis. In the field, black scoriaceous flows, either aphanitic or showing phenocrysts of pyroxene and (or) olivine, were called basalt. Chemical analyses showed that some of these flows were andesite. Lighter colored rocks were commonly called dacite if the phenocrysts were principally plagioclase and rhyolite if quartz was abundant. On the basis of chemical analyses, rocks with less than 54 percent silica were called basalt, rocks with 54 to 68 percent silica were called intermediate, and those with more than 68 percent silica were called rhyolite. Intermediate rocks were divided into basaltic andesite (less than 58 percent silica), andesite (silica content between 58 and 62 percent), and dacite (more than 62 percent silica).

K-Ar Mineral Ages

Because most of the volcanic rocks of the state cannot be correlated on the basis of fossiliferous strata, our stratigraphic designations are based almost entirely on K-Ar dating (Table 9; App. 7).

Radiometric ages are in agreement with relative stratigraphic ages assigned by field relations. The exceptions are where the rocks selected in stratigraphic sequence are so similar in age that the imprecision of the method (the ± value) is greater than the difference of the assigned ages. Figure 26 shows that volcanic activity in the state has continued for the past 19 m.y. In terms of volume, the volcanic outpourings of the gulf have been minor during the past 7 m.y.

Chemical Analyses

Chemical analyses of the volcanic rocks of Baja California (Table 10) show a range of 47 to 74 percent SiO_2 with similarly large ranges for the other major oxides. From a comparison of $(K_2O + Na_2O)$ to SiO_2 (Fig. 27), some generalizations

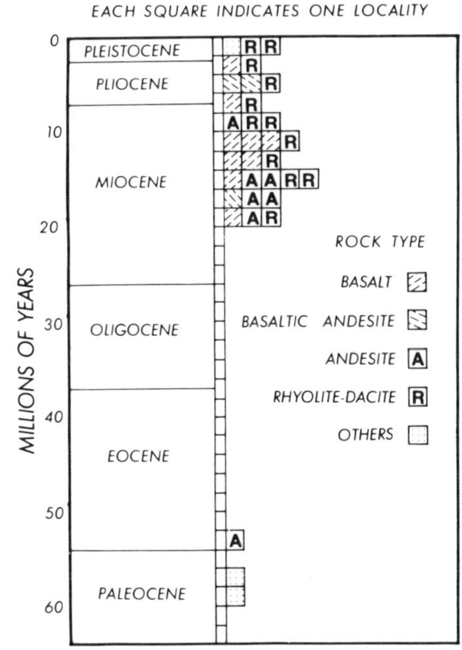

Figure 26. Histogram of ages of Cenozoic volcanic rocks in the state of Baja California. In addition to those of Table 9, dates reported by Hawkins (1970a), Barnard (1968b), and Rossetter (App. 1) are included.

can be made: (1) Nearly all of the rocks fall above the line that Kuno (1969) used to distinguish the alkalic series from the high-alumina series (line 2 in Fig. 27), and all but one lie well above his pigeonitic rock series (line 1 in Fig. 27). (2) Both the alkalic and more calc-alkalic rocks have a wide range in age. (3) There is no simple relation between ($K_2O + Na_2O$) index and either age or spacial

TABLE 10.
CHEMICAL ANALYSIS OF TERTIARY VOLCANIC ROCKS

Sample*	SiO_2	Al_2O_3	Fe_2O_3	FeO	MgO	CaO	Na_2O	K_2O	TiO_2	MnO	P_2O_5	H_2O	CO_2	Total	Analyst
GROUP I. Eocene rock															
524	50.59	12.05	5.79	0.72	8.18	8.08	2.15	5.45	1.53	0.09	1.31	2.24	2.13	100.31	5
GROUP II. Miocene basalt															
B8B-15a	47.12	15.41	..	..	..	9.84	4.13	0.19	1.35	..	..	..	..	..	4
B9B-15	47.72	15.03	9.03	..	11.40	9.68	2.39	1.68	1.14	0.11	..	..	..	..	1
553	49.73	15.19	5.98	2.71	8.94	10.11	3.61	0.63	0.88	0.15	0.20	1.75	0.53	100.41	5
B3A-33	49.58	15.30	..	..	..	9.00	2.97	0.21	1.08	..	..	..	..	..	4
GROUP III. Rocks of Intermediate Composition - Miocene and Pliocene															
B8B-67	55.58	15.41	6.58	..	7.50	7.58	3.03	1.84	0.79	0.08	..	..	..	..	1
523	55.98	16.93	5.66	0.30	3.73	5.35	3.63	4.66	0.88	0.08	0.37	1.67	0.20	99.44	5
B4P-23	56.09	16.91	..	..	..	6.90	4.02	0.70	0.64	..	..	..	..	..	4
B9B-55	57.32	15.48	..	..	..	7.22	3.34	1.43	0.95	..	..	..	..	..	4
517	57.84	15.48	2.95	6.46	2.09	6.07	4.23	1.35	1.59	0.15	0.49	0.65	0.35	99.70	5
531	58.15	18.24	4.78	1.64	2.32	5.98	3.90	1.97	0.89	0.11	0.22	1.07	0.20	99.47	5
B6G-11	58.17	15.65	5.79	..	3.64	6.94	3.39	2/36	0.94	0.07	..	..	..	..	1
HQ-129	58.35	17.38	2.05	4.07	3.52	6.44	3.28	2.50	1.37	0.10	0.20	0.80	..	100.06	2
541	61.31	16.29	2.08	1.89	2.37	5.46	4.35	1.11	0.68	0.07	0.18	3.00	0.38	99.17	5
539	61.94	14.45	4.84	1.82	1.81	4.11	4.01	2.20	1.71	0.07	0.33	1.92	0.23	99.44	5
B5F-85	62.21	16.71	..	..	..	4.56	4.78	1.37	1.05	..	..	..	..	..	4
540	63.02	15.53	3.74	0.48	2.45	4.89	5.19	1.71	0.69	0.07	0.18	1.79	0.35	100.09	5
561	64.30	13.70	1.33	0.25	0.77	2.85	2.90	3.12	0.23	0.04	0.04	8.70	0.82	99.05	5
HQ-134	65.68	14.71	1.12	3.30	1.58	3.12	3.52	2.95	0.67	0.04	0.00	3.03	..	99.72	2
HQ-133	67.50	14.27	1.05	2.05	1.02	2.59	3.37	3.36	0.58	0.04	0.03	4.23	..	100.09	2
GROUP IV. Miocene rhyolite															
559	70.96	12.72	3.28	0.25	0.20	1.92	5.32	3.45	0.39	0.09	0.06	0.84	0.60	100.08	4
518	71.26	13.19	1.80	0.25	0.44	1.58	4.14	4.93	0.30	0.12	0.06	1.28	0.22	99.57	4
550	73.87	12.96	0.79	0.20	0.15	1.24	4.35	4.60	0.15	0.02	0.05	0.91	0.20	99.49	4
GROUP V. Pliocene-Holocene basalt and basaltic andesite of the peninsula															
AOW-1123A	46.51	14.97	6.27	6.63	11.05	8.40	2.20	0.74	0.36	0.96	0.52	1.47	..	100.08	3
B6G-49	46.57	15.70	..	..	..	8.42	3.75	1.80	2.82	..	..	..	..	..	4
B4F-62	48.67	16.22	..	..	..	8.90	3.56	0.52	2.89	..	..	..	..	..	4
AOW-1126C	49.43	16.36	5.67	5.60	8.31	8.64	3.24	0.87	0.22	1.19	0.21	0.12	..	99.86	3
B8B-66	50.56	15.53	8.25	..	9.19	8.87	2.56	2.11	1.16	0.10	..	..	..	..	1
B3G-141	50.79	16.52	8.80	2.30	4.66	7.85	3.84	1.56	1.98	0.12	0.56	1.71	..	100.69	4
552	51.16	13.20	5.63	0.92	8.88	7.86	2.82	3.93	1.55	0.10	0.95	2.45	0.47	99.92	5
B5G-158	51.83	15.04	..	..	..	10.76	3.64	2.76	2.69	..	..	..	..	..	4
C-I, II (avg)	52.05	23.13	8.51	1.75	2.24	3.30	3.10	5.37	0.15	..	0.20	1.16	..	100.96	6
B5W-100b	54.40	15.76	..	..	..	8.60	4.33	1.81	1.14	..	..	..	..	..	4
GROUP VI. Pleistocene basalt of the rift															
Isla Raza	52.01	15.55	9.04	..	5.67	7.07	0.86	1.57	0.39	5.66	..	..	..	97.82	1

1. John Minch (personal commun., 1970)
2. Hirschi and de Quervain (1927 - 1933)
3. K. Willman (Woodford, 1928)
4. Japan Analytical Chemistry Research Institute, Tokyo
5. Michel Delaloye, Institut de Mineralogie, Univ. Geneva
6. K. von Chrustschoff (1885)

NOTE: for additional published analyses of rocks within this province see Larsen (1948) and Hawkins (1970a).
* For location and petrographic description of volcanic rocks, see Appendix 7.

distribution. (4) As compared to other Pacific border areas, the Baja California rocks lie to the alkalic side of the Cascade suites (Kuno, 1969) but correspond to the compositional distribution for Chile and Peru (Pichler and Zeil, 1969). For purposes of discussion, we have divided the analyses into six groups (as shown on Fig. 27):

Group I. The lower Tertiary rock sample was alkalic augite basalt (KA-524).

Group II. Miocene basalt from both sides of the peninsula forms a tight cluster on the $(K_2O + Na_2O)/SiO_2$ diagram but varies widely in K_2O content. The small number of analyses available shows no consistent differences between rocks of the Pacific Coast and Gulf of California provinces. The 4 analyses reported here (Table 11) resemble 14 analyses for Pleistocene basalt from Oregon and Idaho (Leeman and Rogers, 1970, p. 9). The Miocene basalt from the peninsula appears more "tholeiitic" than the Pliocene-Pleistocene basalt and basaltic andesite and shows less primitive strontium-isotope ratios (Table 12; Leeman and others, 1972).

Group III. The intermediate rocks represent most of the volcanism in the gulf depression during the past 8 m.y. The 14 analyses reported here range in composition from 55.6 to 67.5 percent silica and include five basaltic andesites, four andesites, and five dacites. The group straddles the boundary between Kuno's (1969) alkalic

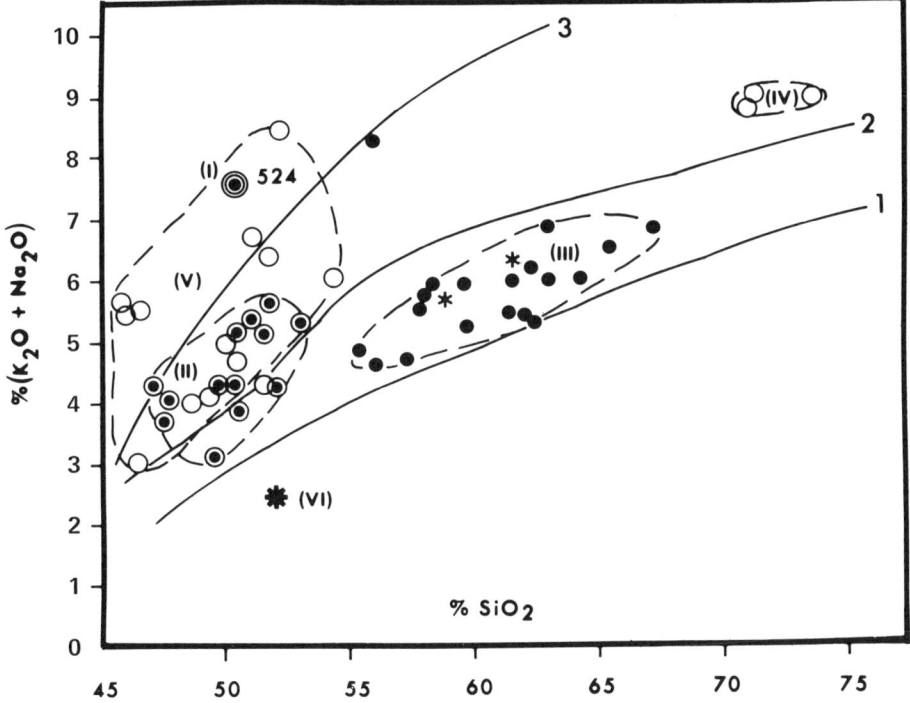

Figure 27. $(K_2O + Na_2O)/SiO_2$ variation diagram for Cenozoic volcanic rocks: line 1, upper boundary of the pigeonitic rock series (Kuno, 1969); line 2, upper boundary of the high-alumina series (Kuno, 1969); line 3, upper boundary of calc-alkalic series (Pichler and Zeil, 1969). The points are grouped as in Table 10 with additional points from Larsen (1948) and Hawkins (1970a). I, Eocene rock (bull's-eye); II, Miocene basalt (dotted circles); III, Miocene-Pliocene rocks of intermediate composition (dots = rocks from the Gulf of California volcanic province, and asterisks = rocks from the Pacific Coast volcanic province); IV, Miocene rhyolites (open circles); V, Pliocene-Pleistocene basalt and basaltic andesite from the stable peninsula (open circles); VI, Pleistocene basalt, Isla Raza (bold asterisk).

series and high-alumina series. Na_2O is generally high relative to K_2O. All are low in Mg, and some have high Fe/Mg ratios.

One sample of Miocene andesite (KA-523), from near Bahía de las Animas) contains more than 8 percent ($K_2O + Na_2O$) in sharp contrast to the other rocks of group III.

Group IV. The three samples of Miocene rhyolite from Baja California range in age from 20 m.y. at Santa Gertrudis to 8.3 m.y. west of Puertecitos, but they plot together on the ($K_2O + Na_2O)/SiO_2$ diagram.

The rhyolite contains balanced weights of Na_2O and K_2O as do the rhyolite domes from the Salton Sea (P. T. Robinson, 1970, written commun.) and rhyolite from the Red Sea rift system (Mohr, 1970).

Groups V and VI. Pliocene and Pleistocene rocks with the field name "basalt" (black, scoriaceous, largely aphanitic) are widespread from the islands of the gulf across the peninsula to the islands of the Pacific borderland. In thin section, these rocks show microlites of intermediate to calcic plagioclase and clinopyroxene. Some show olivine; others, biotite and potassium feldspar.

From their seemingly random distribution, these rocks supposedly would have similar compositions throughout the peninsula; however, the stable peninsular association (that is, from the central, tectonically rigid portion of the peninsula [Chaps. 8, 10]; group V of Fig. 27) is alkalic. All except one of the samples analyzed during this project have more than 5 percent ($Na_2O + K_2O$), and all but one have less than 52 percent SiO_2. This group of rocks resembles the late Cenozoic alkalic olivine basalt described by Leeman and Rogers (1970) from the Basin and Range province. In contrast, most of the younger "basalt" samples from the Gulf of California depression proved to be andesitic in composition and are included in group III.

Group VI contains the one sample of young basalt analyzed from the gulf; it is from Isla Raza (loc. 29) astride the Guaymas lineament. It is low in Na and contains 5.66 percent MnO. It is the only analysis that falls within Kuno's (1969) pigeonitic rock series.

One observation emerges clearly: the magmas of the Cenozoic volcanism were more alkalic than those that formed the Mesozoic batholiths (Fig. 28).

Gulf of California Volcanic Province

In Figure 24, the mid-Cenozoic volcanic rocks of the state are divided into the Pacific Coast volcanic province and the Gulf of California volcanic province. The Gulf of California volcanic province is divided into five subprovinces on the basis of geographic separation. The degree to which they have different chemical or volcanigenic histories is as yet unknown.

Santa Gertrudis Subprovince. The oldest volcanic rock in the Santa Gertrudis subprovince is the rhyolitic tuff breccia (19.8 ± 0.5 m.y. B.P.) that was quarried to build Misión Santa Gertrudis. It is overlain by more than 100 m of andesite that thickens eastward. The andesite is capped by Pliocene or early Pleistocene basalt that is separated from the underlying strata by a major erosional unconformity and locally even rests directly on the basement rock.

San Borja-Punta Animas Subprovince. North of the El Arco-San Francisquito road, the volcanic section provides an almost continuous cover. These rocks extend north to the Sierra la Asamblea (sometimes called Sierra Calamajué, Sierra Yubay, or Sierra San Luis), west to the Pacific coastal plain, and east across Isla Angel de la Guarda (Fig. 24).

TABLE 11.
COMPARISON OF FOUR MIOCENE BASALTS OF BAJA CALIFORNIA WITH PLIOCENE-PLEISTOCENE BASALTS OF OREGON AND IDAHO

	Oregon-Idaho [*] %	Baja California [†] %
SiO_2	46.6	48.00 ± 1.17
TiO_2	1.82	1.11 ± 0.17
Al_2O_3	15.9	15.23 ± 0.14
CaO	10.86	9.34 ± 0.46
Na_2O	2.74	3.27 ± 0.69
K_2O	0.50	0.70 ± 0.62

[*] From Leeman and Rogers, 1970.
[†] One standard deviation.

TABLE 12.
STRONTIUM ISOTOPE RATIOS FOR VOLCANIC ROCKS FROM THE STATE OF BAJA CALIFORNIA

Sample [*]	Rock Type	Age (m.y.)	$^{87}Sr/^{86}Sr$ initial [†][‡]
B5W-100B	basaltic andesite	Holocene (?)	0.7037 ± 3
B4P-23	basaltic andesite	Late Pleistocene	0.7035 ± 3
B6G-49	alkali-olivine basalt	Late Pleistocene	0.7028 ± 3
B6G-62	spinel lherzolite	Late Pleistocene	0.7031 ± 5
KA-552	basalt	2.6 ± 0.5	0.7037 ± 3
KA-538	basalt	4.3 ± 2.0	0.7039 ± 3
KA-531	andesite	5.1 ± 0.4	0.7037 ± 2
B6G-2	basalt	Plio-Pleistocene	0.7029 ± 2
B4F-62	basalt	Plio-Pleistocene	0.7027 ± 3
KA-517	basaltic andesite	7.6 ± 0.4	0.7043 ± 3
KA-541	andesite	8.9 ± 0.6	0.7059 ± 3
B9B-55	basaltic andesite	Pliocene	0.7048 ± 3
KA-527	basalt	10.5 ± 0.4	0.7054 ± 3
KA-526	basalt	11.9 ± 0.5	0.7049 ± 3
KA-549	basalt	13.6 ± 0.4	0.7040 ± 3
B8B-67	basalt	Miocene	0.7048 ± 3
B9B-15	basalt	Miocene	0.7057 ± 2
KA-535	basaltic andesite	16.3 ± 0.5	0.7047 ± 2
KA-524	alkalic basalt	59 ± 1.8	0.7053 ± 2

[*] For location and petrographic description, see App. 7.
[†] Standardized to 0.7080 in E A $SrCO_3$ standard.
[‡] Error is 99.5% confidence limit on single analyses, last significant digit only (actually x 10^{-4}).
NOTE: data courtesy of William Leeman, University of Oregon.

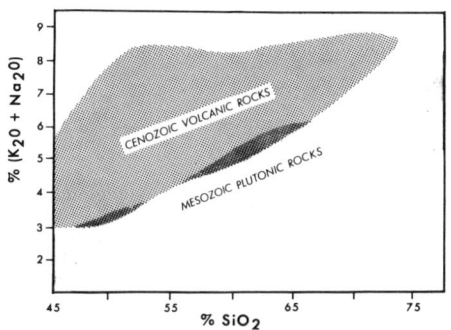

Figure 28. Comparison of $(K_2O + Na_2O)/SiO_2$ ratios for Cenozoic volcanic and Mesozoic plutonic rocks from the Peninsular Ranges.

South of Bahía de las Animas, conglomerate containing large andesite boulders that were dated at 17 m.y. B.P. (KA-523; Table 9) rests on Eocene volcanic rock (loc. 31). The andesite boulders may be equivalent to andesite of similar age at Santa Gertrudis (loc. 35), El Mármol, and the Sierra de los Cucapas. The conglomerate is overlain more or less conformably by basalt and rhyolite with ages ranging from 10 to 14 m.y. The basalt and rhyolite are unconformably overlain by basalt that is 5 m.y. old or less. In the western portion of the area, where the Cenozoic strata are nearly flat lying, the sequence consists primarily of rhyolitic tuff breccia as much as 100 m thick, overlain by basalt. Near Misión San Borja (loc. 28), dating of the rhyolitic tuff from two separate localities yielded ages of 14.0 and 14.2 m.y. (KA-546 and KA-554B). The overlying basalt yielded an age of 12.0 m.y. (KA-553). In the Sierra Sandia area (loc. 34), a similar volcanic sequence unconformably buries andesite mountains that may be the cores of old eruptive centers. Elsewhere, even within the thick Las Animas basin section, rhyolite probably no older than 14 m.y. rests directly on the basement rocks. Therefore, as Beal (1948) suggested, the eruptive centers of mid-Miocene andesite probably were near the crest of the peninsula west of the gulf depression. Furthermore, by implication, a major erosional interval occurred between 17 and 14 m.y. ago.

The apparently thick sections of tilted rock and the variety of volcanic units in the Las Animas-Angel de la Guarda area suggest that this is a section comparable in thickness to those measured by McFall (1968) at Bahía Concepción and by McEldowney (1970) in the Sierra Pinta, which are in excess of 4,000 m.

Puertecitos Subprovince. The oldest volcanic rocks in the Puertecitos subprovince are probably the cores of old andesite centers such as Pico Matomí (loc. 17; Fig. 29), which stands on the crest of the peninsula. Hornblende andesite (KA-535), collected just south of El Mármol (loc. 23), was dated at 16 m.y. B.P. Overlying these older andesite units are two sequences that are primarily pyroclastic rhyolite.

The older rhyolitic sequence is steeply tilted in many places and stands vertically at the south end of Valle de San Felipe (loc. 16). It includes units of columnar-jointed welded tuff, spherulitic vitrophyre, and a variety of volcanic breccia and finer grained pyroclastic rocks. Samples KA-555, KA-556, KA-557, and KA-559, collected from northwest of Puertecitos (loc. 19), were dated (Table 9). Although the dates are not in stratigraphic sequence, their confidence limits overlap, and an age of about 8 or 9 m.y. is indicated. The thickness of the sequence has not been measured but must be at least 2,000 m.

The younger rhyolitic sequence is nearly flat lying and caps the area inland from Puertecitos. The capping strata are only 100 m thick and include layers of massive aphanitic rhyolite, fine to coarse pyroclastic rhyolite, pumice, and vitrophyric rocks. Obsidian from the mesa east of Pico Matomí (loc. 17) was used extensively for points and tools by the pre-Spanish inhabitants. Locally, as at the south end of Valle de San Felipe (loc. 16), wedges of sandstone and conglomerate lie at the base of the younger sequence. Northwest of Puertecitos (loc. 19), two K-Ar ages were obtained from the younger volcanic rocks (KA-560 and KA-561; Table 9) that indicated ages of 6 and 3 m.y. in proper sequence.

No volcanic rocks are known from the higher portions of the Sierra San Pedro

Figure 29. Oblique aerial view southwest across the Matomí volcanic plateau. Along Arroyo Matomí (right foreground), a tilted sequence of Miocene rhyolite is overlain by flat-lying Pliocene sedimentary and rhyolitic strata. In the left foreground, a hill of Miocene rhyolite protrudes above the depositional level of the Pliocene strata. Pico Matomí, a middle Miocene andesite center (upper right), and Cerro San Juan de Dios, a monadnock of Mesozoic volcanic rocks (upper left), also protrude above the flat-lying volcanic cover. At the far right, the Pliocene strata lap onto the granitic rocks (G) of the Sierra San Pedro Mártir. Photograph by John Shelton.

Mártir, although a large block of andesite was found in one of the steep tributaries leading into Valle San Felipe (Robert Slyker, App. 1). Similarly, the Sierra San Felipe is largely devoid of volcanic rocks, but volcanic strata are included in the downdropped blocks between and within the desert ranges of the gulf depression.

Southern Sierra Juárez-Sierra Pinta Subprovince. North from the Sierra San Pedro Mártir and Sierra San Felipe, the volcanic section thickens. It is possible to travel from rhyolite mesas surrounding Santa Catarina in the southern Sierra Juárez (loc. 9), across the Sierra Tinaja (loc. 10) and Sierra Pinta (loc. 11), to the tidal flats of the Gulf of California and to see almost nothing but volcanic strata. These volcanic ranges, close to the international border and easily accessible, have gone almost entirely unmentioned in the literature and are not shown on previous geologic maps of the peninsula.

Overlying the basement in the southern Sierra Juárez is a discontinuous basal conglomerate unit that contains volcanic and basement-rock clasts derived from the east. Biotite from an andesite clast in the conglomerate has an age of 57.8 ± 1.5 m.y. (KA-563; Table 9). The section over the conglomerate consists of four separate volcanic units (in ascending order): widespread andesitic lahar breccia; discontinuous basalt; mesa-capping, dense, and locally welded rhyolitic tuff; and discontinuous basalt. The basalt overlying the lahar breccia yielded an age of 13.6 ± 0.4 m.y. (KA-549); the mesa-capping rhyolitic tuff, an age of 10.0 ± 0.5 m.y. (KA-550). The entire volcanic sequence is probably not more than 300 m thick at any point on the stable Sierra Juárez. East of the main gulf escarpment, however, is an area of north-trending fault blocks of tilted strata that are thicker and consist of a larger variety of volcanic units. There are massive flows of columnar basalt more than 100 m thick, lahar deposits, water-laid pumice breccia, and rhyolite units composed of spherulites ranging from pepper or pea size to good "potato beds." Vitrophyric rock textures include perlite, some with obsidian nodules, but little true obsidian useful for fashioning tools.

In the San Felipe quadrangle that adjoins the Sierra Pinta on the south (Andersen, 1969, App. 1), the volcanic exposures are too discontinuous to construct a composite stratigraphic section. In the southern Sierra Pinta (James, App. 1; loc. 11), volcanic rocks 1,500 m thick rest on the high-relief surface of the basement rock. In the northern Sierra Pinta (McEldowney, 1970, App. 1; loc. 11a), a section (Fig. 30) almost 4,500 m thick rests on basement rock at only one point. The steeply tilted strata and the many unconformities within the thick sequence suggest that this area was the center of both tectonic and volcanic activitity. Ages of 7.6 ± 0.4 to 9.5 ± 1.0 m.y. (Table 9) were determined for basalt (KA-540), basaltic andesite (KA-541), and dacite (KA-517) from the Sierra Pinta area.

Jacumba-Imperial Valley Subprovince. North from the Sierra Pinta and the southern Sierra Juárez, volcanic rocks are absent except in the area within a few kilometers of the international border east of Jacumba, San Diego County (loc. 3). The volcanic rock of the border area consists of andesite breccia less than 100 m thick overlain by olivine basalt. Miller (1935a) named these the Jacumba Volcanics and named the conglomerate that they overlie the Table Mountain Conglomerate. The highest exposures of basalt cap two tiny buttes south of La Rumorosa at an elevation of 1,500 m (loc. 4). From these isolated remnants, the unit descends via fault steps on both sides of the international border almost to sea level adjacent to the Laguna Salada in Baja California and to the foot of the Coyote Mountains in Imperial County. The basalt at Jacumba was dated (Hawkins, 1970a) at 18.7 ± 1.3 m.y. B.P. The underlying lahar andesite yielded ages of 18.6 ± 0.8 and

18.5 ± 0.9 m.y. (J. Minch and D. Krummenacher, 1971, written commun.). A block of the probably correlative Alverson Andesite from the Vallecitos area (SE1/2, sec. 15, T. 14 S., R. 9 E., Imperial County) yielded a K-Ar age of 20.4 m.y. (R. H. Merriam and T. Downs, 1970, written commun.).

In the Sierra de los Cucapas (loc. 5), Barnard (1968b, p. 73-79) described andesite 500 m thick that he named the Colonia Progresso volcanics. Although these Tertiary volcanic rocks cover only approximately 5 km^2, Barnard reported an abundance of volcanic dikes; this suggests that Tertiary volcanic rocks were once more extensive. The 15.3 ± 0.8 m.y. whole-rock age (Barnard, 1968b) from dacite in the Colonia Progresso volcanics corresponds to ages of Miocene andesite in the southern subprovinces.

Younger volcanic rocks are also present in this northern area: the isolated Cerro Prieto (loc. 6) 25 km south of Mexicali is rhyodacite (Robinson and Elders, 1971). Farther north is the Pleistocene(?) Truckhaven Rhyolite and the late Pleistocene (possibly 16,000 yr B.P.) Obsidian Buttes at the southern end of the Salton Sea (Muffler and White, 1969).

Summary of the Gulf of California Volcanic Province. Although it appears that in each of the five subprovinces the complexity and aggregate thickness of the volcanic section increases toward the gulf, it remains to be established whether the axis of volcanic activity has been under the gulf. Certainly, the evidence of most recent volcanic activity is the thermal areas around Obsidian Buttes north of the border, the thermal area adjacent to Cerro Prieto (loc. 6), and the late Pleistocene eruptions on the small volcanic islands along the axis of the Guaymas lineament (loc. 24).

Centers of andesitic eruption exist in the Sierra de la Giganta south of Bahía Concepción, in the Sierra Sandia (loc. 34), near San Borja (loc. 28), and south of the Sierra San Pedro Mártir (loc. 17), all within the ridged backbone of the peninsula. This fact does not, of course, establish that the axis of earlier Miocene volcanism was in this area because even larger early Miocene centers may be buried beneath younger Miocene volcanic rocks closer to or beneath the gulf.

The existence of rhyolite that yielded an age as young as 3 m.y. suggests that volcanic rocks should be found interbedded with Pliocene marine strata in the state of Baja California, but so far none has been discovered. They do occur in the Pliocene rocks at Santa Rosalía and north of Loreto in the territory of Baja California. This probably means that Pliocene and younger volcanic rocks are limited to the part of the gulf that is now submerged. Perhaps in the Oligocene and Miocene Epochs, volcanism was distributed among many rotating blocks; whereas in late Cenozoic time, it had become concentrated on central rifts.

Pacific Coast Volcanic Province

Miocene and younger basalt units are known to occur at many points on the continental borderland and adjacent Pacific coast (Fig. 24). The best documented of these is the Rosarito Beach Formation. Andesite such as that of the Cerro Jesús-María area (loc. 2) is uncommon.

Rosarito Beach Formation. Minch (1967) defined the Rosarito Beach Formation as a sequence of five members with a total exposed section less than 400 m thick (Fig. 31). The Mira al Mar Member is the lowest; it is a sedimentary breccia in which Minch (1967) found a meager fauna that indicated a Miocene age. Table 13 (from Minch, 1967) shows cobble counts indicating that the Mira al Mar Member

TABLE 13.
COBBLE COUNTS IN THE MIRA AL MAR
MEMBER OF THE ROSARITO BEACH FORMATION

Rock Type	Northern sandy breccia (130 Counts) %	Southern sandy breccia (111 Counts) %	Earthy breccia (121 Counts) %
Serpentinite	22.3	14.4	10.7
Glaucophane schist	18.5	20.7	26.4
Glaucophane quartz schist	11.5	6.3	12.4
Bedded chert	17.6	4.5	7.4
Quartz-albite	3.1	11.7	4.1
Miscellaneous schist	3.8	3.6	9.9
Saussuritized gabbro	7.6	8.1	5.8
Gabbro	3.1	13.5	1.7
Granitic rock	3.1	6.4	7.4
Rhyolite	4.2	4.5	6.6
Quartzite	3.1	3.6	2.5
Graywacke	2.0	2.7	5.0
Totals	99.9%	100.0%	99.9%

NOTE: from Minch, 1967.

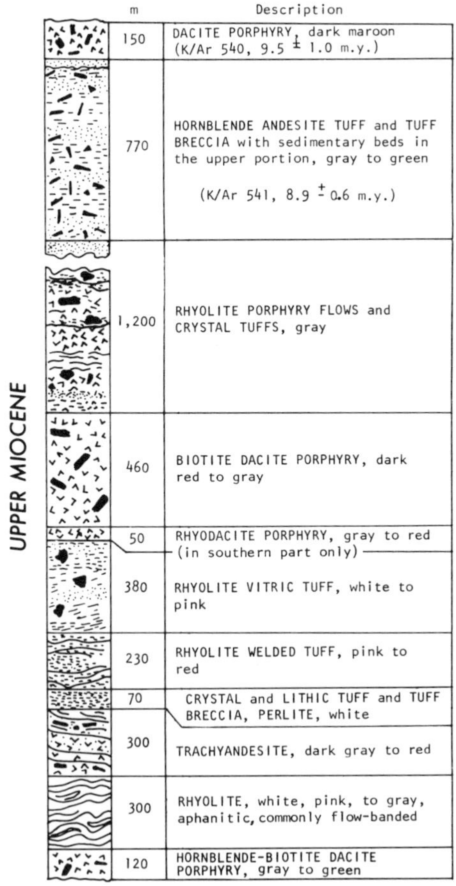

Figure 30. Stratigraphic section of volcanic rocks in the northern Sierra Pinta (loc. 11a, Fig. 24; from McEldowney, App. 1). Base of section is an unconformity. K-Ar localities are shown on Plate 1A.

m	Description
150	DACITE PORPHYRY, dark maroon (K/Ar 540, 9.5 ± 1.0 m.y.)
770	HORNBLENDE ANDESITE TUFF and TUFF BRECCIA with sedimentary beds in the upper portion, gray to green (K/Ar 541, 8.9 ± 0.6 m.y.)
1,200	RHYOLITE PORPHYRY FLOWS and CRYSTAL TUFFS, gray
460	BIOTITE DACITE PORPHYRY, dark red to gray
50	RHYODACITE PORPHYRY, gray to red (in southern part only)
380	RHYOLITE VITRIC TUFF, white to pink
230	RHYOLITE WELDED TUFF, pink to red
70	CRYSTAL and LITHIC TUFF and TUFF BRECCIA, PERLITE, white
300	TRACHYANDESITE, dark gray to red
300	RHYOLITE, white, pink, to gray, aphanitic, commonly flow-banded
120	HORNBLENDE-BIOTITE DACITE PORPHYRY, gray to green

(UPPER MIOCENE)

is mainly derived from offshore Franciscan-type basement rocks, which are believed to have been exposed to the west during Miocene time.

The four upper members of the Rosarito Beach Formation at its type locality are predominantly basalt with interbedded tuffaceous sandstone and shale. The modes (Table 14) of basalt from these members indicate olivine basalt with no systematic change through time. Near the type locality, basalt in the Costa Azul Member was dated by the K-Ar method at 14.3 ± 2.6 m.y. B.P. (Hawkins, 1970a).

This work was extended by Schulte (App. 2), Pendarvis (App. 2), and Hofman (App. 2). The correlation between the type section (loc. 1; Minch, 1967) and the area north of La Misión (loc. 7) is shown in Figure 31. In the La Misión area, the volcanic rocks near the top of the section are more siliceous than the basalt of the four upper members of the Rosarito Beach Formation at its type locality; therefore, the La Misión rocks are classed as basaltic andesite. Plagioclase from the basaltic andesite near La Misión was dated at 16.1 ± 2.1 m.y. B.P. using the K-Ar method (KA-539; Table 9).

Overlying the basaltic andesite in the La Misión area, Schulte (App. 2) discovered a sedimentary sequence composed of an interbedded sandstone and tuff unit that grades upward into a marine sandstone unit. The marine sandstone unit is well exposed at the southwest end of Mesa de los Indios (N. 50° E. from Rancho

TABLE 14. MODES OF BASALT FROM THE ROSARITO BEACH FORMATION

	Costa Azul Member middle basalt	Costa Azul Member upper unit	Amado Nervo Member (eastern block)	Las Glorias Member	Los Buenos Member
Plagioclase	60.8 An66-26	50.3 An62-31	42.5 An68-35	57.0 An68-35	54.2 An66-11
Augite	20.8	20.4	27.5	16.5	24.4
Olivine	5.8	7.8	11.5	9.5	6.4
Magnetite	5.4	8.1	7.0	9.5	4.2
Glass and Alteration	7.2	11.0	11.5	7.5	10.8

Note: After Minch, 1967.

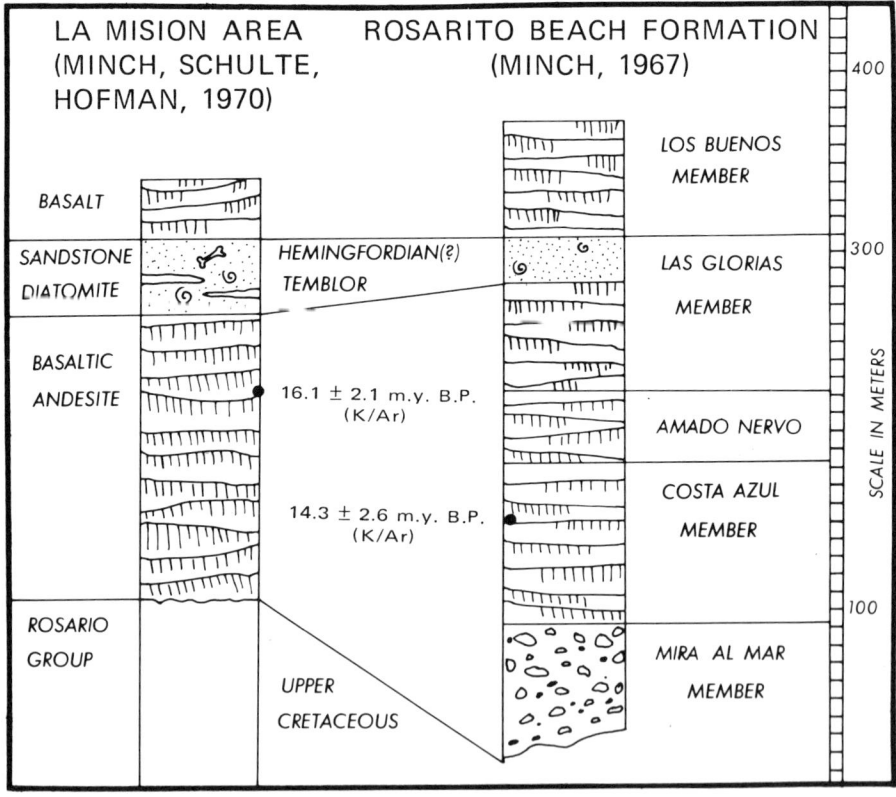

Figure 31. Correlation of the Miocene sections of La Misión and Rosarito Beach. Hemingfordian and Temblor are the vertebrate and invertebrate stages, respectively.

la Pila). Here, the marine sandstone unit consists of (in ascending order) 5 m of fossiliferous (*Chione temblorensis* and *Turritella ocoyana* Conrad), tuffaceous, feldspathic arenite; 10 m of well-indurated, fine-grained, angular, fossiliferous (also *Chione temblorensis* and *Turritella ocoyana* Conrad), feldspathic arenite; 4 m of diatomite (with diatoms and radiolarians); and 6 m of poorly indurated, fine-grained, angular, feldspathic arenite (with *Orbulina* sp., sharks' teeth, and bones and teeth of fish and marine mammals [a faunal list is given by Minch and others, 1970]).

The sedimentary sequence just described is overlain by an upper breccia unit as much as 30 m thick. To the west, the breccia caps the marine sandstone unit; to the east, the breccia caps the lower interbedded sandstone and tuff unit and, still farther east, laps onto the basement rocks.

West and north of the area described by Schulte (App. 2), Pendarvis (App. 2) and Hofman (App. 2) also recognized this sedimentary sequence. Hofman discovered 30 m of basalt overlying Schulte's marine sandstone unit and suggested the correlation of this basalt with the Los Buenos Member of the Rosarito Beach Formation (as shown in Fig. 31).

Minch (1967) summarized the similarities of the Rosarito Beach Formation, the San Onofre Breccia (Woodford, 1925), and other Miocene volcanic and sedimentary rocks of the continental margin. During the Miocene Epoch, part of the now submerged Franciscan-like basement rocks of the continental borderland must have stood above sea level to the west of Rosarito Beach. The Vizcaíno Peninsula and Isla Cedros are remnants of this Miocene ridge. During the Miocene Epoch, the depression between the ridge of Franciscan-like rocks and the interior of the peninsula was alternately occupied by shallow seaways and freshwater valleys. Basalt eruptions occurred throughout the continental borderland (Emery, 1960; Moore, 1969; Merifield and others, 1971).

Jesús-María Andesite. East of Tijuana (loc. 2), between the international border and the railroad, is a cluster of andesite domes. The largest is Cerro Jesús-María, which rises 739 m above the coastal terrace. These domes are composed mostly of aphanitic rocks of low color index. The surface on which the andesite rests antedates the Pleistocene Otay Mesa. Similar rocks are found as intrusive necks 30 km to the south (that is, immediately east of Cerro Coronel) and in the northeast corner of Valle de las Palmas to the southeast (Frazer, App. 2). If there are correlative rocks, they might include Cerro Calaveras, which is a dome of columnar dacite east of Carlsbad, and Morro Hill east of Fallbrook. Both Cerro Calaveras and Morro Hill are in San Diego County. The rocks in San Diego County are briefly mentioned by Larsen (1948, p. 108-111). The rocks in Cerro Calaveras can only be dated as post-Eocene. No analogous post-Eocene but pre-Pliocene volcanic deposits have been observed elsewhere in the Pacific Coast volcanic province of southern or Baja California.

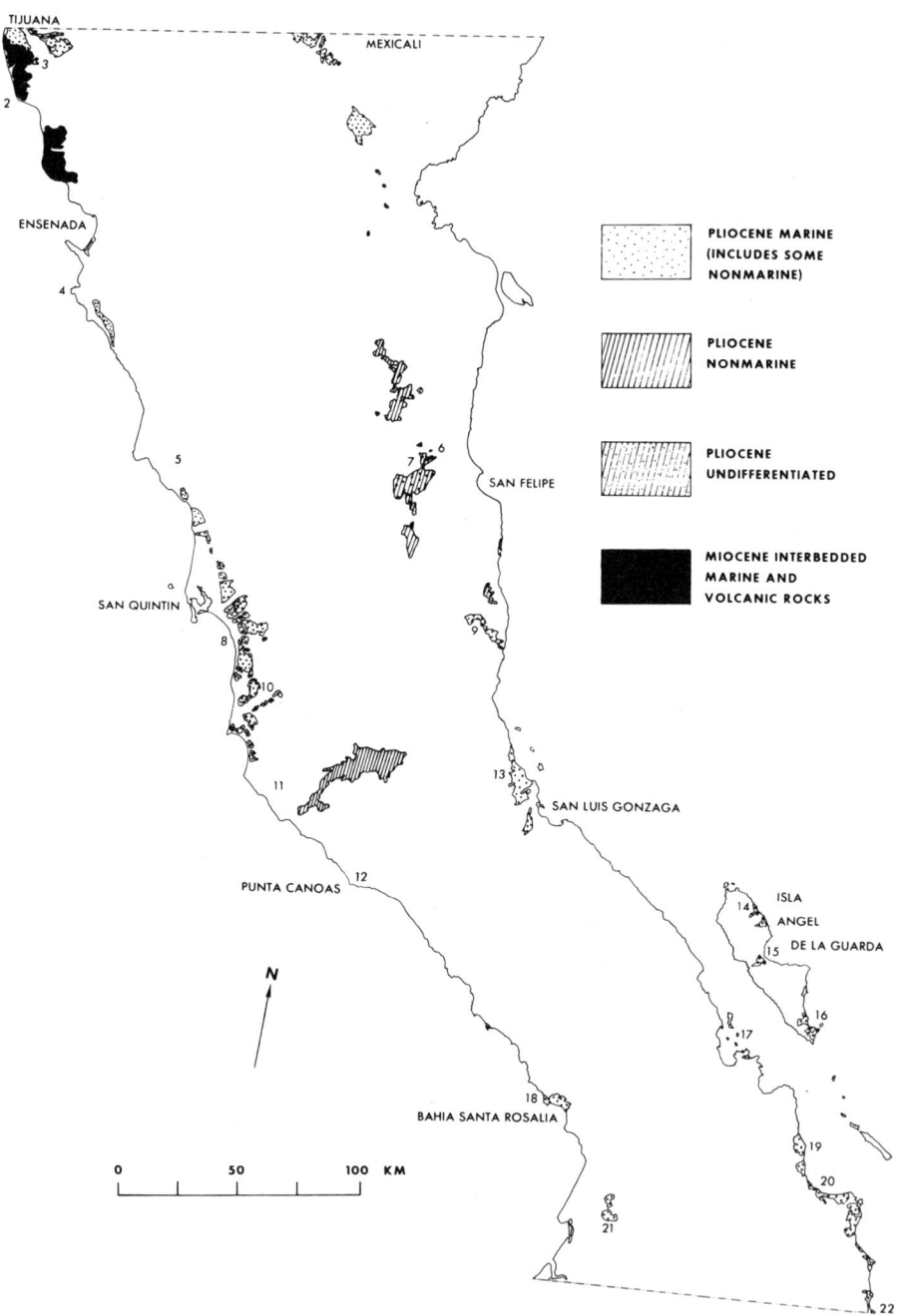

Figure 32. Distribution of Pliocene and Miocene sedimentary rocks. Numbers 1 through 22 correspond to localities cited in the text.

6

Flooding of the Gulf of California and the Continental Borderland

PLIOCENE EPOCH ON THE PACIFIC COAST

By Pliocene time, the Baja California Peninsula had assumed much of its present geographic form. The Gulf of California extended as far north as the Transverse Ranges in California, U.S.A., and the Pacific coastline stood close to its modern position. To the west, the continental borderland had subsided into a series of linear troughs and ridges (Moore, 1969, p. 44) with a few islands remaining exposed. The peninsula extended south at least to the La Paz embayment.

During early Miocene time, the sediment derived from the peninsula was transported westward past the present shoreline and deposited on a broad, open continental terrace that is now the continental borderland (Moore, 1969). The mid-Miocene orogeny folded and faulted this area, and Pliocene faulting formed new basins.

Marine Pliocene deposits are found along the Pacific coast almost continuously from the Los Angeles basin in southern California to lat 28° N. In Baja California, most of these strata are preserved in relatively thin marine terrace deposits that extend only a few kilometers inland to well-preserved shorelines. In several places, the marine Pliocene rocks appear to be continuous with nonmarine deposits that extend farther inland. Such deposits are preserved in the mesa south of Tijuana (loc. 3 in Fig. 32, which shows locations for Chap. 6; Minch, 1967), in the San Telmo area (loc 5), and inland from Bahía Santa Rosalía (loc. 18; Fife, App. 1).

Marine Pliocene strata north of the international border are called the San Diego Formation (Dall, 1898). South of the border, the general usage has been Formación Salada (for Rancho Salada at the southern end of the Magdalena Plain; Heim, 1922); however, Santillán and Barrera (1930) adopted the name Formación Contil Costero for the strata north of El Rosario (loc. 10), and Mina (1956, 1957) adopted Formación Almejas for rocks of the Vizcaíno Peninsula. Mina (1957, p. 206) also adopted the name Formación Bandeada for 105 m of strata between Tijuana and Ensenada; this apparently encompassed the combined Miocene-Pliocene section in that area.

Minch (1967) extended the name San Diego Formation to the Playas de Tijuana-Rosarito Beach area (locs. 1 and 2). Formación Salada was designated as a group by Durham (1950) and in Mexican publications has come to mean all marine Pliocene strata of the peninsula. Formación Cantil Costero (Santillán and Barrera, 1930) was designated as middle and upper Pliocene, whereas the San Diego Formation was interpreted as upper Pliocene (Valentine and Rowland, 1969).

The marine Pliocene deposits of southern California and northern Baja California are littoral sandstone and conglomerate derived from the basement and older sedimentary strata immediately to the east. The fauna from these deposits was described and interpreted by Hertlein and Grant (1939), Gunther (App. 2), Minch (1967), Andersen (App. 2), and Valentine and Rowland (1969). In the territory of Baja California, the fauna was described by Darton (1921, p. 747), Heim (1922, p. 544-546), Hertlein (1925), Beal (1948, p. 80-81), and Mina (1957, p. 179, 207-208).

Although the Pliocene beds of western Baja California are nearly horizontal, their elevations vary. The highest Pliocene shoreline in the area southeast of Playas de Tijuana is about 200 m above sea level (loc. 3). From Rosarito Beach (loc. 2) to Punta Banda (loc. 4), the highest Pliocene terrace is apparently below current sea level. From Punta Banda south to Socorro (loc. 8), extensive Pliocene deposits occur at elevations of 100 to 200 m. From Socorro southward, the Pliocene terrace climbs to elevations of around 300 m northeast of El Rosario (loc. 10). Pliocene strata are 500 m above sea level at Mesa San Carlos (loc. 11), 100 to 200 m along the coast south of Punta Canoas (loc. 12), and at less than 50 m above sea level farther south (locs. 18 and 21). Orme (1971) reported that the highest terrace is at 345 m at Punta Banda, at 357 m at El Rosario, and below sea level at Camalú about halfway between Punta Banda and El Rosario. On paleontologic evidence, the maximum terrace elevations to which Orme referred are of Pliocene age (Acosta, App. 1).

EARLIEST MARINE STRATA IN THE GULF OF CALIFORNIA DEPRESSION

The oldest confirmed marine strata in the northwestern portion of the Gulf of California basin are diatomite units that underlie the Imperial Formation northwest of San Felipe (loc. 6; Andersen, App. 1). This horizon has been dated as late Miocene (Mohnian-Delmontian) on the basis of pelagic foraminifera (Andersen, 1969). Other foraminifera, radiolarians, and diatoms from the same material tend to confirm this identification (Ingle, 1974; Mandra and Mandra, 1972).

King (1939, p. 1693) reported an account of a deep well at Empalme near Guaymas in the state of Sonora that penetrated "a considerable thickness of marine Tertiary strata." During the 1940 cruise of the *E. W. Scripps*, Durham (1950) made an unsuccessful search for Tertiary marine outcrops in the Empalme area. Recently, many deep water wells have penetrated Tertiary marine strata in southern Sonora including the Empalme area, and the microfauna from these wells was reported (Gómez, 1971) as of upper, lower, and (possibly) middle Miocene age.

PLIOCENE ROCKS OF THE GULF OF CALIFORNIA DEPRESSION

The recognition of marine strata of Pliocene age in the Gulf of California province goes back to the survey by Blake (1858). The earliest paleontological reports on the fauna of Imperial Valley, California, were by Orcutt (1889), Fairbanks (1893), and Vaughn (1900). Early reports of the deposits along the gulf were by Gabb (1882), Fuchs (1886), Dall (1898), Arnold (1906), and Hanna and Hertlein (1927).

The stratigraphic treatment of the Imperial Valley deposits began with Kew (1914), followed by G. D. Hanna (1926), Woodring (1931), Tarbet and Holman (1944), and Dibblee (1954). Recent discussions are by Woodard (1974) and Stump (App. 1).

G. D. Hanna (1926) gave the name Imperial Formation to the coral reefs exposed in Alverson Canyon on the south side of Coyote Mountains, Imperial County, California. This name is used for all marine strata above the Split Mountain Formation (Dibblee, 1954). The overlying nonmarine strata containing petrified wood and vertebrate remains are generally known as the Palm Spring Formation named for Palm Spring on Vallecitos Creek, 2 km west of the old Vallecitos stage station (Woodring, 1931). Durham (1950) attempted to correlate the name Imperial Formation as used in the Imperial Valley with the Pliocene formation names used in Baja California (Fig. 25).

Despite an abundant fauna, the time-stratigraphic assignment of the marine gulf deposits has been a topic of persistent disagreement. Dickerson (1918), Bramkamp (1935), and Tarbet and Holman (1944) identified the Imperial fauna as Miocene; Vaughn (1917), Grant and Gale (1931), and Durham (1950) determined that the fauna was Pliocene. Dibblee (1954) indicated "upper Miocene or possibly lower Pliocene." In recent years, there has even been a tendency to call the fauna Pliocene to Pleistocene (Allison, 1964), but studies of vertebrates (Downs and White, 1968) placed the lower part of the overlying Palm Spring Formation in the upper Pliocene. This would seem to limit at least those portions of the Imperial Formation that underlie the Palm Spring Formation to the Pliocene Epoch or earlier. Some marine strata of the gulf depression that have been referred to the Imperial Formation are possibly the marine equivalent of the Palm Spring Formation. The problem of paleontologic dating arises from the fact that the warm-water faunas of the gulf cannot be compared directly with the faunas of the Pacific coast of southern and Baja California.

The thickest exposed section of Pliocene rocks in the peninsula occurs just north of Loreto where Beal (1948, p. 78) reported as much as 1,000 m of folded strata unconformably overlying the Comondú Formation. In the Imperial Valley of California, U.S.A., as much as 1,500 m of marine Pliocene strata were reported (Dutcher and others, 1972, their Fig. 5).

The only other area in which marine Pliocene rocks measure more than a few hundred meters is the area east of Valle San Felipe (loc. 7) where marine fossils are interbedded with coarse talus and gritty sand of granitic provenance. The folded Valle San Felipe strata crop out as much as 700 m above sea level (Andersen, 1969, App. 1). K-Ar dating of volcanic rocks in the Puertecitos area implies that the last major folding of the gulf basin deposits occurred between 6 and 8 m.y. B.P.

Wilson and Rocha Moreno (1955) reported a number of tuff beds in lower Pliocene strata at Santa Rosalía. North of the territory of Baja California, no direct evidence of volcanism contemporaneous with marine deposition has been found, even though volcanic activity appears to have continued throughout the Pliocene Epoch (see K-Ar ages in Table 9).

Nearly horizontal terraces (Fig. 33) containing fossiliferous marine strata are at elevations close to current sea level north of Puertecitos (loc. 9), just north of Bahía San Luis Gonzaga (loc. 13), near Bahía de los Angeles (loc. 17), on the coast east of Llano San Pedro (loc. 19), on adjacent gulf islands (locs. 14, 15, and 16), and along the coast from Bahía San Rafael (loc. 20) south to lat 28° N. (loc. 22; Fig. 33). These terrace deposits are designated as marine Pliocene (Tpm) on the geologic map (Pls. 1-A, 1-B, and 1-C), but it is doubtful that they are the same age as the folded strata near Valle San Felipe (loc. 7). For example, a few kilometers north of Puertecitos, horizontal beds of marine strata crop out near the coast. To the west, they abut (or are faulted against) a sequence of volcanic strata from 3 to 6 m.y. in age. Nowhere was the volcanic rock observed overlying the marine strata.

Marine deposits that may correlate with the Imperial Formation or with the Salada Group are found as far north as the foothills of the Transverse Ranges, along the Colorado River (the Bouse Formation), and in basins of the southeastern Mojave Desert (Smith, 1970) of southern California. Pliocene deposits are reported in the El Gulfo area of the state of Sonora (Merriam, 1965); farther south in Sonora, however, they are unknown.

There are some lithologic similarities in the widespread occurrences of marine Pliocene strata. At or near the base, there are magnesian limestone beds that are in places algal or coquinal and as much as 100 m of gypsum (or halite in the most northeasterly occurrences). Most of the overlying strata are sandstone and conglomerate with occasional yellow claystone and shell reefs. Toward the top of the section, nonmarine facies predominate.

A series of fault-bounded basins apparently occupied the western part of the gulf depression during Pliocene time. The inflow of sea water into the basins may have been balanced by evaporation; this allowed some of the more distant and shallower basins to dry up periodically, thus producing salt and gypsum deposits. Owing to the pattern of interior drainage, many basins may have received little detrital sediment, explaining those in which organic or chemical deposits rest directly on an irregular bedrock surface. Gradually, the basins filled with clastic debris, to the point where the depositional surface was above sea level. Alternatively, the transition to nonmarine conditions could have been caused by the formation of detrital sills excluding the sea; such a sill exists today south of Mexicali. Interior basins below sea level may have been many times alternately flooded and dried during the past 10 m.y.

Figure 33. Marine terraces of the Gulf of California. Oblique aerial view (facing northwest) of Bahía San Francisquito. In the foreground, Pliocene-Pleistocene marine terraces are cut in granitic and metamorphic rock. The highest terrace is approximately 100 m in elevation.

PENINSULAR INTERIOR DURING LATE CENOZOIC TIME

By Pliocene time, the early Tertiary surface on which the Miocene volcanic rocks were deposited had been considerably elevated and extensively eroded. In places, interior basins were filled with fluvial and lacustrine deposits. Most of these deposits cannot be dated precisely. In several localities, they are capped by Pliocene volcanic rocks. Near Rosarito, Fife (App. 1) distinguished Miocene, Pliocene, and Pleistocene basalt on the basis of stratigraphic and geomorphic relations. The Miocene basalt was dated at 12 m.y. B.P. (KA-553; Table 9), and an ultra-alkalic Pleistocene basalt, as 2.6 m.y. B.P. (KA-552; Table 9). The intermediate basalt appears on geomorphic grounds to be closer in age to the later date and is therefore almost certainly Pliocene.

At Santa Gertrudis near lat 28° N., the capping basalt flows rest with distinct unconformity on both the basement rocks and the Miocene volcanic sequence. A camalid bone of uncertain age was found in fluvial deposits that postdate the basalt units (Don Savage, 1965, written commun.).

CONCLUSIONS

The time at which the sea first entered the Gulf of California depression cannot now be determined other than to say that the earliest time is no later than late Miocene (Mohnian-Delmontian) on the peninsular side and earlier on the Sonoran side. A more widespread flooding occurred in the Pliocene. Perhaps the sea entered more than once. Inevitably, the most complete record of the sea is to be found in the central basins now submerged beneath the gulf.

Of course, the entry of marine water does not tell us when the deep basins of the gulf opened. When did the elevational disparity between the ranges or plateaus of the peninsula and the Gulf of California depression first arise? Did the early and mid-Miocene volcanic fields accumulate at elevations of 1,000 to 2,000 m and then subside, or was the entire area now occupied by the gulf and peninsula near sea level in Miocene time and then the peninsula rose?

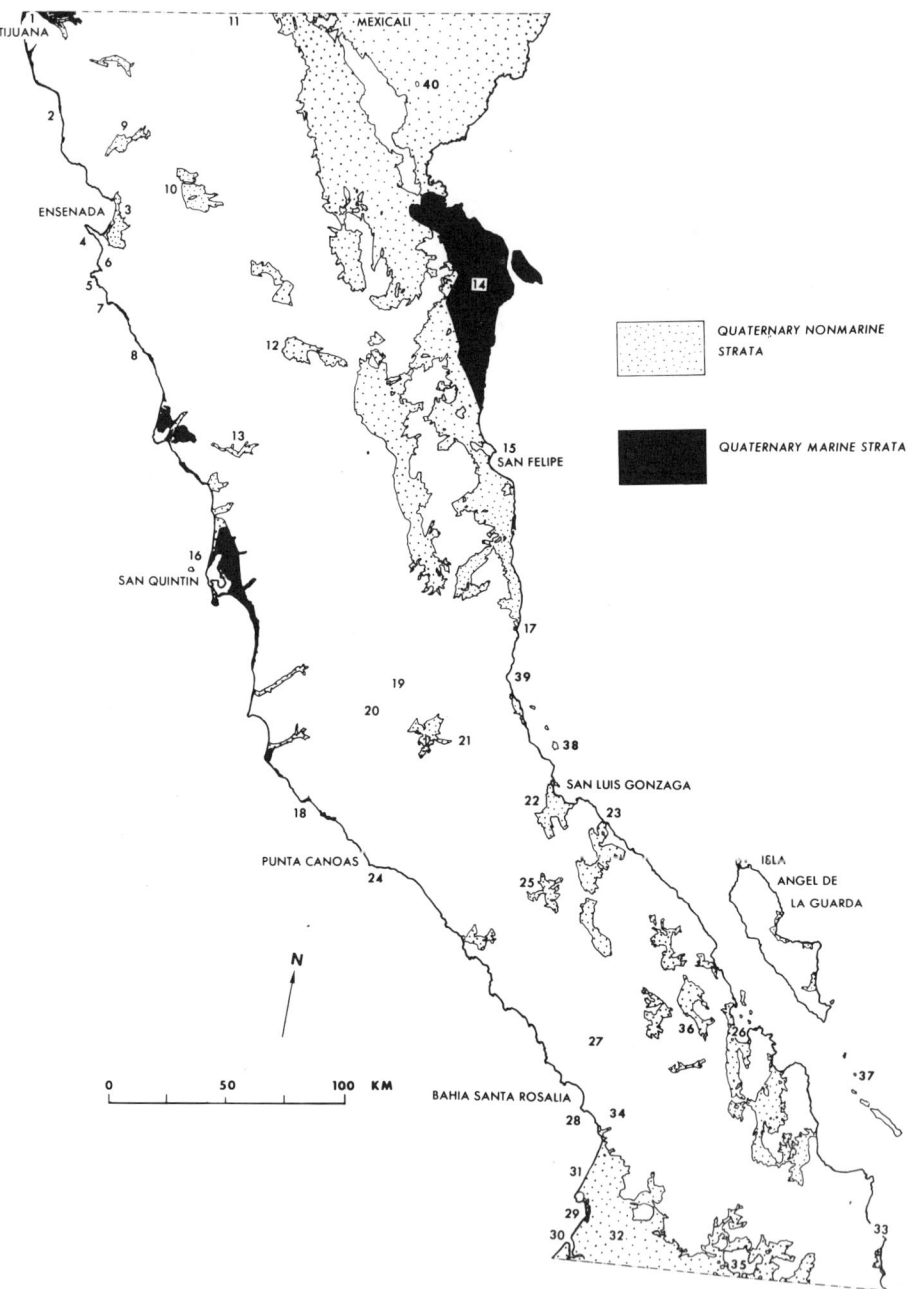

Figure 34. Distribution of Quaternary sedimentary rocks. Numbers 1 through 40 correspond to localities cited in the text.

7

Pleistocene and Holocene Epochs

PACIFIC COASTAL TERRACES

Just north of the international border, in the San Diego area, a number of Pleistocene terraces have been recognized (Hertlein and Grant, 1944), and several formation names have been applied to the deposits capping them. Of these, the two most distinctive units are the red ferruginous conglomerate and sandstone that cap the mesa at elevations of 100 and 150 m and the very fossiliferous sandstone or coquina that is found near sea level. The mesa-capping rocks are referred to as the Lindavista terrace deposit (M. A. Hanna, 1926); the sea-level deposits are named the Bay Point Formation by Hertlein and Grant (1939) for the occurrence on the north shore of Mission Bay. South of the border, formal formation names have not been devised, but a similar red layer caps the hills south of Tijuana and on South Island of Islas los Coronados. A fossiliferous terrace deposit a few meters above sea level extends almost continuously from Playas de Tijuana to La Misión (locs. 1 to 2 in Fig. 34, which shows locations for Chap. 7).

Schroeder (App. 1) described the Pleistocene terrace deposits just south of Ensenada (loc. 3), and Acosta (1970) described those extending from Punta Banda (loc. 4) to Punta China (loc. 5). Emerson (1956, 1960) investigated the deposits from north of Punta Santo Tomás (loc. 6) to Punta San José (loc. 7). Emerson and Acosta applied the name Punta China terrace to the prominent bench standing between 5 and 16 m above sea level. Lindgren (1888), Allen and others (1961), and Acosta (App. 1) recognized as many as 12 higher terraces in the Punta Banda area (loc. 4), north of the Agua Blanca fault (Pl. 3); the highest terrace was almost 350 m above sea level. Acosta also reported that the highest terrace (250 m) south of the Agua Blanca fault was dated as Pliocene on the basis of fossils collected by John Minch (1966, written commun.).

Emerson (1960) described the terrace at Punta Cabra (loc. 8). The terrace at San Quintín (loc. 16) was studied by Orcutt (1921), Jordan (1926), Woodford (1928), and Gorsline and Stewart (1962). From San Quintín to Punta Canoas (loc. 24), our field parties mapped a succession of four terraces. The highest of these contains Pliocene fossils; the lowest is the prominent Pleistocene terrace referred to above (terrace T1 in Fig. 35).

It may be significant that, in those areas where the coastline projects westward into the Pacific, the Pliocene and older Pleistocene terraces are tilted seaward and attain greater elevations, whereas in all areas the youngest terrace is essentially horizontal. This strongly suggests that the elevated areas were produced by post-T2, pre-T1 deformation that produced broad buckles on axes subnormal to the coast.

On terrace T4 (Fig. 35), fossiliferous Pliocene strata are overlain by basalt. Terraces T2 and T3 are capped by a thin cobble layer that includes basalt. Farther north, supposedly correlative surfaces are of marine origin. Terrace T1 forms a prominent, nearly horizontal bench 10 to 20 m above sea level. The marine deposits of T1 fill irregularities on a channeled surface, suggesting that sea level fell below its present elevation between formation of T2 and T1. According to Emerson and Addicott (1958), the low terraces (loc. 18) were deposited under essentially the same conditions that exist along this coast today.

South from Punta Canoas, fossiliferous Pleistocene beds occur just above sea level. In many places, they are covered by fluvial gravel or eolian sand. Emerson and Hertlein (1960) studied a late Pleistocene marine fauna in a terrace about 7 m above sea level at Punta Rosarito (loc. 28; lat 28°40' N.).

Fife (App. 1) described the Bahía Santa Rosalía localities (lat 28°30' to 29° N.) as follows:

Near the coast there are several terraces. The most prominent are at the 15-20, 60, and 140-foot elevations. The fifteen to twenty-foot terrace affects only sedimentary rocks; it consists of a 1,000-foot wide bench south of El Muertito; and can readily be recognized [to the] south in the Puerto Santo Domingo Quadrangle. . . . The higher terraces have been cut up to several hundred feet into the granitic rocks north of El Tomatal. North and south of the mouth of Arroyo Rosarito the sixty-foot and the 140-foot terraces merge, obscuring the sixty-foot terrace. The sixty-foot terrace follows the present coastline around the mouth of the arroyo, while the 140-foot terrace curves upstream for a few miles.

East of Santa Rosalillita the higher surface rises as a gentle plain for about four miles, reaching a maximum elevation of about 200 feet. Beyond this elevation the terrace or terraces have been so modified by erosion that one terrace grades imperceptibly into another. This is about 100-200 feet lower than the highest surface preserved by the Miocene volcanic rocks.

South from lat 28°30' N., the coastal plain becomes broad and submergent. Beneath the sand dunes, the entire coastal terrace is capped by hard calichified

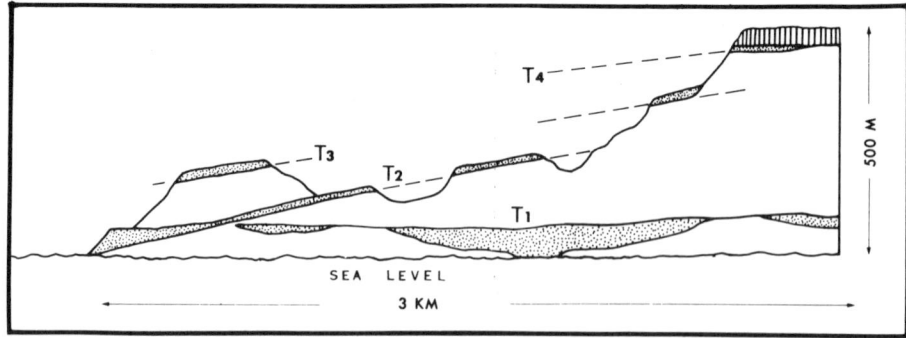

Figure 35. Diagrammatic sketch of terraces T1 through T4 west of Mesa San Carlos, Pacific coast of Baja California, at lat 29°30' N.

white sandstone as much as 3 m thick. This overlies reddish-brown, moderately friable sandstone and argillaceous siltstone that continues downward at least to the bottom of the deepest arroyos.

NONMARINE COASTAL DEPOSITS

Dunes (Cooper, 1967) and cobble berms along the modern coast form barriers across the mouths of valleys and fringe coastal plains. The barriers, such as those south of lat 28°30' N. (loc. 31), create basins for nonmarine deposition.

South of Playas de Tijuana, Katsuo Nishikawa (1968, oral commun.) of the Universidad Autónoma de Baja California, Ensenada, excavated skeletons of elephant and horse. Nonmarine gravel underlies and overlies the upper Pliocene or lower Pleistocene basalt of Mesa San Carlos. Andersen (App. 2) described nonmarine Pleistocene deposits as much as 33 m thick near lat 28°30' N.

Nearshore nonmarine deposits underlie most of the valleys and low coastal areas. In Valle El Rosario, the banks of the inner arroyo expose many layers of midden shells in the old valley fill.

INTERIOR DEPOSITS

The lack of fossil evidence makes it difficult to determine the age of deposits in the interior valleys, but the intermittent lake basins presumably contain Pleistocene to Holocene deposits. A number of these deposits have been cored and studied palynologically by A. Cross (1965, oral commun.) of Michigan State. Arnold (1957) described shore levels on Pleistocene lakes in the Laguna Seca de Chapala. He believed that the peninsula has been appreciably tilted to the west during the past few thousand years. Alluvial deposits cover the Llano del Berrendo (loc. 32), the interior valleys from La Bachada to Arroyo Calamajué (loc. 23), the Laguna Seca de Chapala (loc. 25), the El Mármol-Misión San Fernando area (loc. 21) north to Rancho Cartabón (loc. 19) and to El Arenoso (loc. 20), Valle San Telmo (loc. 13), Valle Trinidad (loc. 12), Valle San Rafael (loc. 10), Valle Guadalupe (loc. 9), the area northwest of La Rumorosa (loc. 11), and many smaller areas. Most of these elevated valleys are covered by relatively shallow deposits of alluvium and are analogous to such valleys as El Cajon, Poway, Santa Maria (also called Ramona), and Perris in the Peninsular Ranges north of the international border.

Wittich (1909) and Böse and Wittich (1913, p. 357-359) described deposits of very recent marine shells at elevations as high as 2,000 m. Darton (1921) and Beal (1948, p. 117-119) not only repeated this account but attempted to add collaborating observations. Both authors repeated the figure "3,500 feet" (1,000 m), choosing to overlook similar accounts of submergence to 1,800 and 2,000 m below sea level.

The original report is reproduced in its entirety (App. 8), because we are convinced that all of these "very recent" shell deposits were manmade. This midden material is mainly restricted to the Pacific drainage of the peninsula and is found only locally along the shore of the gulf. The abundance of shells diminishes rapidly inland from the Pacific coast. The prehistoric residents apparently transported edible shellfish for more than a day's journey as indicated by the fact that isolated calcareous rubbish heaps are found high in the mountains. Many of these distant deposits consist of a small cluster of shells such as might have been carried by an individual on a single trip.

In short, we see no justification for the often repeated speculation that the peninsula has undergone rapid submergence and re-emergence during late Pleistocene or Holocene time.

GULF OF CALIFORNIA COASTAL DEPOSITS

Marine Pleistocene deposits are described at Santa Rosalía (Wilson, 1948; Wilson and Rocha Moreno, 1955) and are present in narrow nearshore terraces south of El Barril (loc. 33), where they unconformably overlie marine Pliocene strata; around Bahía de los Angeles (loc. 26) and adjacent islands; just north of Bahía San Luis Gonzaga (loc. 22); at Puertecitos (loc. 17); and along the coast near San Felipe (loc. 15; Walker and Thompson, 1968; Andersen, App. 1). Anderson (1950) and Emerson and Hertlein (1964) described marine Pleistocene deposits on several islands in the southern half of the gulf.

The thickest Pleistocene sections were deposited in deltaic, brackish, mud-flat, and saline-basin environments. In Borrego Valley north of the international border, almost 3,000 m of Pleistocene strata have been reported from drill holes (W. S. Kerr, 1949, written commun.). Most of the Pleistocene section is nonmarine. Downs and White (1968) reported a detailed study of the vertebrates in a nonmarine Pleistocene section west of Imperial Valley, California, U.S.A. Much of the sediment that filled the northern half of the gulf depression during the Pleistocene Epoch came from the Colorado River. A study of fossiliferous clasts shows that they were derived from the Grand Canyon of Arizona (Merriam, 1965; Barnard, 1968b). Barnard reported that in some Pleistocene conglomerate units located near the Sierra de los Cucapas, one out of every ten clasts contained fossils that indicated Grand Canyon provenance.

The minor exposures of marine Pleistocene strata and the abundance of nonmarine Pleistocene strata in the basins at the head of the gulf suggest that the northern gulf area was either elevated or barred from marine incursion during much of Pleistocene time.

VOLCANISM

Young volcanic rocks appear to be closely related to the Gulf of California rift system. The rhyolitic Obsidian Buttes of the Salton Sea (Robinson and Elders, 1971) and the rhyodacitic Cerro Prieto (Barnard, 1968b; Robinson and Elders, 1971) on the Colorado River Delta 25 km south of Mexicali (loc. 40) are adjacent to active thermal areas. They appear to lie along the proposed axis of the most northerly of the gulf spreading centers (Elders and others, 1971).

Farther south, the Guaymas lineament (Fig. 55) appears to mark another active structure related to the modern gulf. It strikes into a youthful crater of basaltic andesite at the peninsular coast at Cerro León (loc. 39). The northwest portion of this lineament is marked by a chain of volcanic islands (loc. 38; Fig. 36) of andesitic to rhyolitic composition (Rossetter and Gastil, 1971). The basalt of Isla Raza (loc. 37) may be part of the same volcanic lineament. The volcanic products of the rift include considerable dacite-rhyolite, in contrast to the exclusively mafic Pleistocene to Holocene volcanism elsewhere in the peninsula.

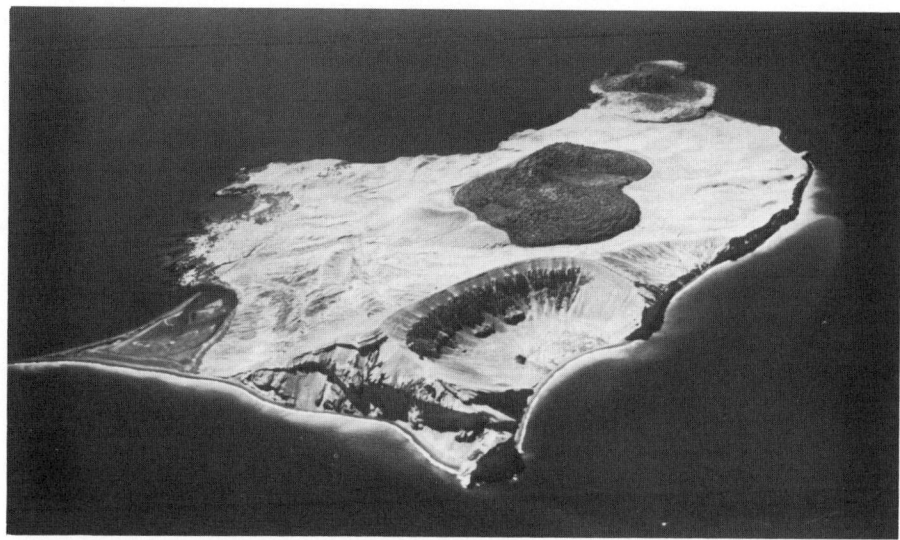

Figure 36. Oblique aerial view to the northwest along the line of eruption of Isla San Luis, largest of the islands north of Bahía San Luis Gonzaga. Miocene andesite is exposed on the western points and in the floor of the breached explosion crater. Most of the island is covered by stratified pumice. Two obsidian domes of Holocene age, upper right, have filled and overflowed explosion craters.

Away from the gulf rifts, Pleistocene to Holocene volcanic activity appears to be a continuation of the predominantly basaltic (including basaltic andesite) volcanism which began in the Pliocene Epoch. Prominent areas are the craters at San Quintín (loc. 16), the extensive field between Jaraquay and Arroyo San José (loc. 27), scattered eruptions along the western foothills south from El Rosarito (loc. 34) to lat 28° N. near El Arco (loc. 35), and the fissure eruption east of Playa Amarga (loc. 36; Fig. 37). Beal (1948, p. 84) described the young eruptions south of lat 28° N.:

The more striking but less widespread younger flows, most numerous near the crest of the range between 25° and 28° N. Lat., have been ejected so recently that streams, obstructed by the flows, have not yet adjusted themselves to the changed topography. Owing to the arid conditions of the area, these lava flows show practically no effects of weathering; apparently very recently the black masses have flowed a short distance over the sparse vegetation and uneven topography of the desert area and cooled with their edges several feet high. These areas present conspicuous and weird features to a landscape of utter desolation. The flows are doubtless Quaternary in age.

The Tres Vírgenes, a classic cluster of volcanic cones rising to an elevation of almost 2,000 m northwest of Santa Rosalía, erupted in the year 1746 and apparently smoked in the year 1857 (Mooser and Rayes, 1961; Ives, 1962).

Gorsline and Stewart (1962) reported that the craters at San Quintín are intermediate in age between the older and younger dune fields and overlie midden deposits dated at 5,000 to 6,000 yr B.P. Gorsline and Stewart believed that the northern cones of the group are the youngest, having been deposited less than 3,000 yr ago. Woodford (1928, p. 345) similarly concluded that "the last activity of the field may have been Recent, even historic." Many other cones and flows on which we have no dates appear to be as fresh as those of San Quintín.

Furthermore, hot springs, carbon dioxide, and sulfurous emanations occur at many places within and adjacent to the peninsula.

SUBMERGED PORTION OF THE CONTINENTAL BORDERLAND

Emery (1960) published an extensive study of the area off southern California. Offshore from the state of Baja California, two detailed studies have been made. Krause (1965) studied the tectonic, bathymetric, and geomagnetic patterns of the continental borderland; Moore (1969) added seismic reflection profiling to delineate the rock units.

Moore concluded that the formation of the modern basins dates from mid-Pliocene time, with tectonic action continuing to the present. These basins are filling with

Figure 37. Oblique aerial view to the east of Pleistocene basalt flows east of Llano de Amarga, northwest of Bahía de los Angeles. The Ballenas Channel is at the upper edge of the photograph. Note that the basalt appears to have issued from a north-trending fracture. In this area, Pliocene and Pleistocene basalt flows rest on granitic rocks exposed by post-Miocene erosion. Photograph by John Shelton.

sediment at rates of 5 to 212 cm/1,000 yr. The present filling has been going on for about 1 m.y. The Quaternary sediment is little disturbed except where cut by recent faults. "Deposition by turbidity currents has been of paramount importance, virtually masking any hemipelagic deposition" (Moore, 1969, p. 2).

PACIFIC COASTAL LAGOONS

Gorsline and Stewart (1962) studied the origin and modern sedimentation of the lagoon at San Quintín (loc. 16). Cooper (1967) discussed the coastal dunes of California, U.S.A., and Baja California as far south as Arroyo Socorro, 30 km south of San Quintín.

Laguna Guerrero Negro (also called Laguna San Miguel; loc. 30), Scammons Lagoon (also called Laguna Ojo de Liebre), and Laguna Manuela (loc. 29) compose the submergent interface between Bahía Vizcaíno and the Baja California syncline of Beal. These lagoons were described by Emery (1957), Stewart (1958), Phleger and Ewing (1959, 1962), Inman and others (1966), and Phleger (1969). The lagoons formed about 6,000 to 7,000 yr B.P., and the present barrier beach, about 1,800 yr B.P. Evidence exists for an earlier lagoonal cycle older than 30,000 yr B.P.

GULF OF CALIFORNIA

Thompson (1968) and Walker and Thompson (1968) described the late Pleistocene to contemporary mud flats (loc. 14) that extend from the mouth of the Colorado River down the west side of the gulf almost to San Felipe. Fine-grained sediment from the Colorado River and evaporite have been accumulating in this area for at least the past 40,000 yr. Walker (1967) described the formation of ancient and modern red beds in the deserts west of the gulf, including the areas studied by Walker and Thompson.

Byrne and Emery (1960) and van Andel (1963, 1964) reported the distribution of modern sediment within the gulf. Cross (1966) described the pollen and spore content and distribution in the gulf sediment.

Van Andel (1964) recognized three sedimentary facies in the northern gulf (his Fig. 43, p. 274). Next to the present shoreline is modern sand of local origin; the sand extends less than 10 km into the gulf. The central, deeper water section (including the Wagner and Delfin basins) is dominated by the sediment of the modern Colorado River. These sediment deposits continue southward into the Salsipuedes basin west of Isla Angel de la Guarda and into the Tiburón basin east of Isla Tiburón. The areas of the northern gulf that are less than 10 m deep are mantled with a layer of post-Pleistocene transgressive sand, deposited during the most recent rise in sea level. The sand has a high glauconite content; this implies a slow rate of clastic deposition.

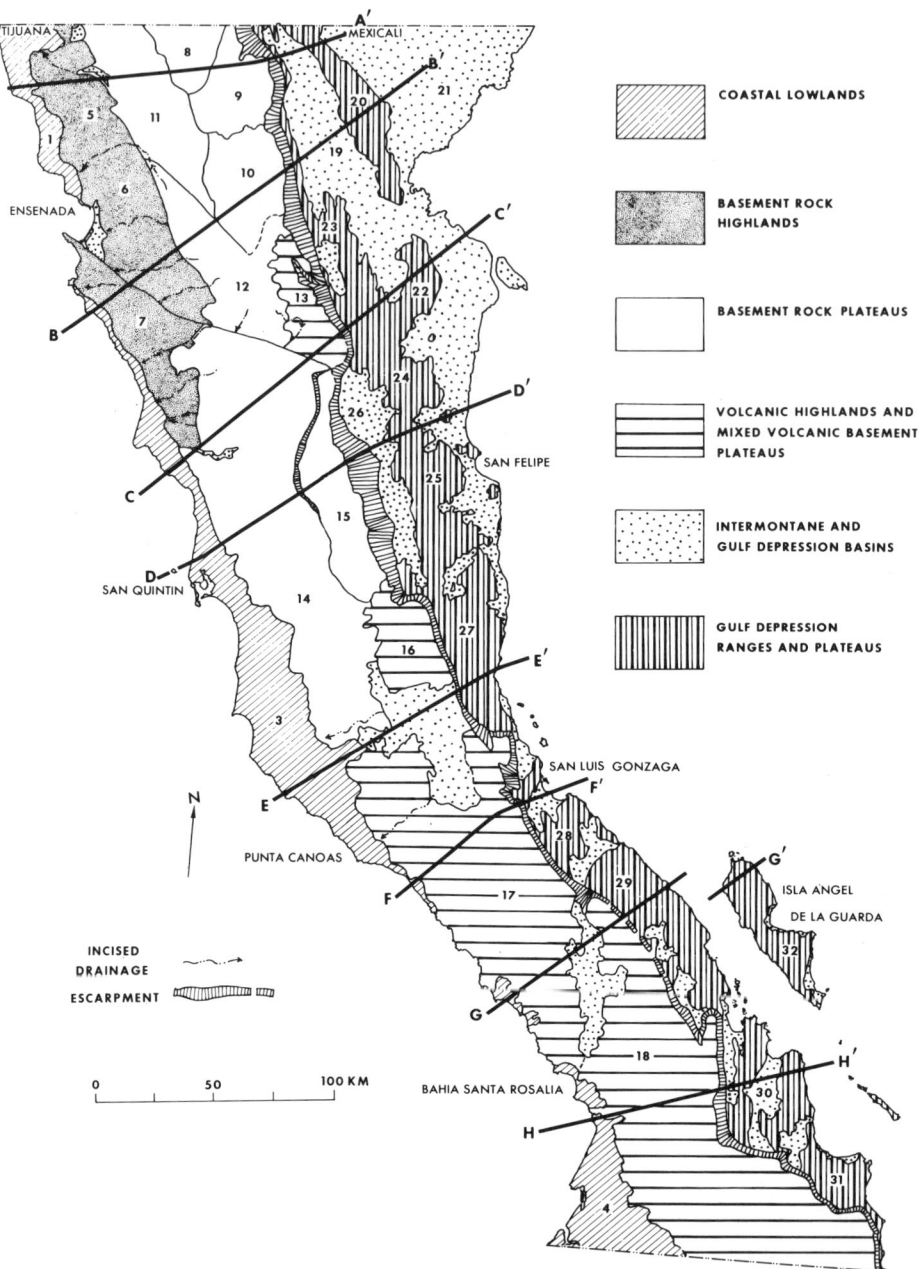

Figure 38. Geomorphic provinces of Baja California. For profiles A-A' to H-H', see Plate 4. Numbers correspond to the following geographic provinces: 1, northwest coastal area; 2, Todos Santos coastal plain; 3, central coastal area; 4, Llano del Berrendo; 5, La Zorra block; 6, Ensenada block; 7, Tomás block; 8, McCain plateau; 9, La Rumorosa plateau; 10, Laguna Hansen surface; 11, Tecate surface; 12, Alamo block; 13, Santa Catarina plateau; 14, Peterson block; 15, Sierra San Pedro Mártir; 16, Matomí plateau; 17, Jaraguay block; 18, San Borja block; 19, Laguna Salada; 20, Sierra de los Cucapas and Sierra del Mayor; 21, Colorado River Delta; 22, Sierra Pinta; 23, Sierra Tinaja; 24, Cerro Borrego block; 25, Sierras San Felipe and Santa Rosa; 26, Valles San Felipe and San Pedro; 27, Puertecitos block; 28, Gonzaga block; 29, Remedios block; 30, Los Angeles-Las Animas block; 31, El Barril block; 32, Isla Angel de la Guarda.

8
Geomorphology

PREVIOUS INVESTIGATIONS

Gabb (1882) recognized that the Baja California Peninsula has a topographic axis and a drainage divide that lies very close to the Gulf of California. A short, steep slope leads down to the gulf; and a long, gentle slope, to the Pacific. Goodyear (1888) and Lindgren (1888), each of whom had visited only the Sierra Juárez, drew attention to the topographic parallels with the Sierra Nevada in California, U.S.A. Emmons and Merrill (1894, p. 491-495) and Merrill (1897) described the peninsula as far south as lat 29°30′ N. and pointed out the broad interior valleys. The historic route from San Ignacio to San Pedro Mártir along the axis of the peninsula crossed first one and then another of these high valleys.

The only pertinent studies devoted specifically to geomorphology are those of Sauer (1929) and Miller (1935b) in San Diego County. Both were concerned with the extensive old erosion surfaces on the high bedrock plateaus. Sauer, a follower of Walter Penck, believed that these surfaces formed simultaneously at several levels. Miller, a follower of William Morris Davis, believed that a single surface had been formed and had subsequently been uplifted differentially by Pliocene faulting. (An excellent discussion of this classic difference between Penck and Davis may be found in vonEngeln, 1942.) The most impressive of the old surfaces (the McCain plateau, described below) can be seen from Interstate 8 between La Posta and Jacumba, California, U.S.A., or between Tecate and La Rumorosa on Mexican Highway 2. This surface is lower than and extends between the Laguna Mountains to the north and the Sierra Juárez to the south.

Beal (1948) discussed the ideas of Sauer and Miller as they applied to Baja California. Beal believed that the "old erosion surfaces" were formed in the Eocene Epoch. There appear to be examples of what both Sauer and Miller described. As shown in Figure 38 (which shows the geomorphic provinces discussed in Chap. 8; topographic profiles are shown in Pl. 4), the boundary between the Laguna Hansen surface and the La Rumorosa plateau may be a product of differential simultaneous erosion; the boundary between the Laguna Hansen surface and the Alamo block, however, is the northwestern extension of the San Miguel fault zone (Shor and Roberts, 1958).

The sea floor adjoining Baja California has been extensively studied. Work in the continental borderland west of southern and Baja California was pioneered by Davidson (1897), who recognized and named most of the prominent submarine canyons. The geomorphic pattern and its structural implications within the continental borderland were studied by Emery and others (1952), Emery (1960), and Uchupi and Emery (1963). Krause (1965) made detailed investigations of the southern borderland, and Moore (1969) added extensive geophysical studies.

The bathymetric pattern of the Gulf of California was interpreted by Shepard (1950), Byrne and Emery (1960), and Rusnak and others (1964). Byrne and Emery (1960) constructed a set of profiles extending from the outer edge of the continental borderland to Arizona and Sonora. Their profiles (Fig. 39) illustrate the general westward slope of western Mexico.

GEOMORPHIC HISTORY

The present topography of the Baja California Peninsula closely reflects its geologic history and structure. The area can be subdivided into four regions of distinct topographic character. These, in turn, can be subdivided into geomorphic provinces (Fig. 38).

Along the west, a line approximating the Santillán y Barrera line (Chaps. 4 and 10) separates the coastal terrace and continental borderland area from the plateaus of the central peninsula. To the north, between the Santillán y Barrera line and the peninsular plateaus, is a region of rugged topography developed primarily on prebatholithic volcanic rock. This topographic region continues into California, U.S.A.

The central plateaus rise eastward to elevations near 2,000 m, east of which they are sharply truncated by the main gulf escarpment. This spectacular topographic escarpment can be followed almost continuously from Mount San Jacinto, Riverside County, south almost to La Paz near the southern end of the peninsula. The escarpment separates the relatively stable peninsular highlands, which we call the "stable peninsula" (all that portion of the peninsula that has acted as a rigid block), from the Gulf of California depression. The gulf depression is divided into basins and ranges, separated by high-angle faults.

Knowledge of the Late Cretaceous topography is fragmentary, but it appears that the Pacific coast of Baja California at that time was close to and similar to the present Pacific coastline. Littoral and neritic Cretaceous strata were deposited against the stacks and sea cliffs of an exceedingly rugged, steep, and generally linear coastline (Acosta, App. 1; McGee, App. 1; Reed, App. 1; Nordstrom, 1970; Peterson and Nordstrom, 1970; Mickey, App. 1). Leading to this coastline were deeply incised canyons, some of which can be seen today still filled with coarse, fluvial debris.

Shallow Paleocene and Eocene seas lapped onto the peninsula, in places beveling the bedrock terrane that lay inland from the Cretaceous coastline. During this interval, the area now occupied by the Peninsular Ranges developed a low, rolling relief—which we call the "old erosion surface"—crossed by streams, many of which flowed in well-defined channels (Minch, 1970). The streams flowed westward, transporting gravel derived from Mesozoic and older terranes in Sonora and southern Arizona. The gravel deposits are found today in isolated patches that preserve remnants of the early Tertiary old erosion surface from which the modern highland

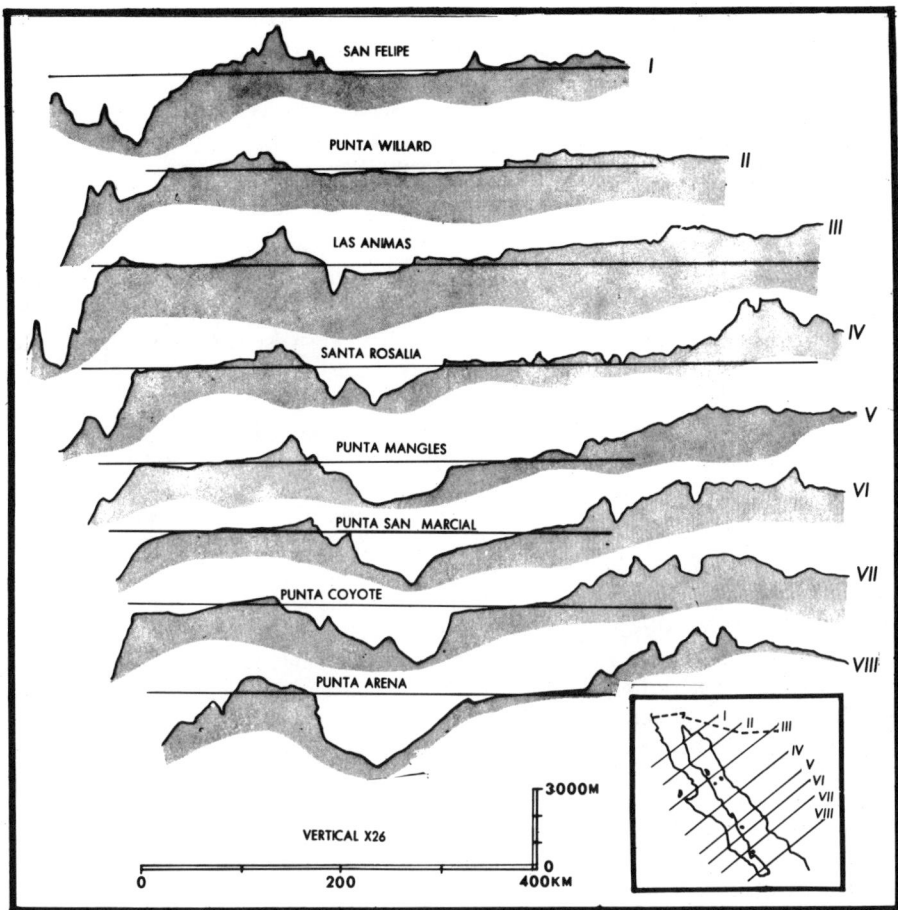

Figure 39. Topographic profiles from the continental escarpment eastward across Baja California, the Gulf of California, and Sonora (from Byrne and Emery, 1960). See also Plate 4.

plateaus have formed. If the old erosion surface is recognized on the basis of the gravel patches still resting on it, it can only be concluded that many of the peaks of today also existed as peaks in Eocene time.

Thick, locally derived deposits of early or pre-Miocene age in the southern part of the state of Baja California suggest an early Tertiary topography of basin-and-range type. These deposits resemble those forming today along the coast northeast of the Gulf of California.

We do not know to what extent the pre-Miocene landforms were once buried by sedimentary and volcanic rocks that have since been stripped away. Scattered remnants of gravel and volcanic rocks suggest that the thick mid-Tertiary deposits that are still found near the gulf thinned toward the Pacific and toward the international boundary. The peninsula probably was low throughout most of the Tertiary Period, and the surface of erosional equilibrium was not greatly out of adjustment with that which existed in the Eocene Epoch.

During Pliocene time and continuing to Holocene time, the extensive bedrock surface with its low relief and veneer of sedimentary and volcanic rocks was uplifted, tilted, and partially stripped of its cover. During uplift, a number of local erosional surfaces formed below the level of the exhumed old erosion surface. Some of these may have formed in equilibrium with temporary stands of sea level, but most appear to be related to resistant ledges formed by inliers of metamorphic rocks within the batholithic terrane.

In the gulf depression, extensive pediments were formed across the uptilted blocks.

PACIFIC COASTAL PROVINCES

From the Coronado escarpment, approximately 15 km west of the present coastline, to near the eastern limit of Late Cretaceous deposition, the geomorphic pattern is directly related to the post-Eocene Pacific coastline. The nature of the western margin is not clear. The 100-fathom (183-m) contour (Moore, 1969) marks the top of a topographic break that can be followed almost continuously from the international border to Bahía Sebastián Vizcaíno. Except near Ensenada, this break is deflected only by the submarine canyons that cross the coastal terrace. Seaward from the international border, the break may be fault controlled, but to the south, this has not been demonstrated.

Since middle Cretaceous time, the Pacific shoreline has been close to its present location (Gastil and Allison, 1966). The easternmost exposure of Upper Cretaceous and lower Tertiary rocks forms a straight line that trends parallel to the coastline and marks the eastern boundary of the Pacific coastal provinces. In some places, late Tertiary coastal terraces extend inland but have not been included in the coastal provinces because they have little effect on the physiography.

The broad marine terraces that characterize the Pacific coastal provinces (Orme, 1971) were cut during late Pliocene and Pleistocene time and may be related to glacial changes in sea level. The terraces have been broadly warped but, except for the Agua Blanca system, are broken only by minor faults. In places, they have been deeply dissected by the drainage channels that cross them. Many of these streams were once graded to a level below present sea level, and their lower courses are filled with fluvial and marine deposits.

Northwest Coastal Area. The northwest coastal area extends from north of Oceanside, San Diego County, to within 10 km of Ensenada. North of the international border, the coastal area is characterized by a series of at least three well-developed terraces (M. A. Hanna, 1926; Peterson, 1907b): the Poway terrace at about 250-m elev, the Lindavista terrace at about 100-m elev, and the La Jolla terrace at 30-m elev. The Lindavista terrace can be followed to the mesa south of Tijuana; M. A. Hanna's (1926, p. 194) La Jolla terrace at Pacific Beach and La Jolla, California, U.S.A., is equivalent to the Nestor terrace named by Ellis (1919, Pl. 6) south of San Diego Bay. A terrace at approximately 30 m fringes the coast to within 20 km of Ensenada and may correspond to the La Jolla terrace. This terrace, nowhere more than 5 km wide, is capped by a nearly continuous layer of Indian midden material (Allen and others, 1961, p. 57).

The inland edge of the northwest coastal area is marked by mesas whose surface is capped by Miocene basalt. This surface reaches an elevation of 703 m on Mesa Redonda but has been incised in places as much as 500 m. Immediately north of La Misión, the coastal mesas slope inland; this suggests a relative uplift of the coastal area.

Todos Santos Coastal Plain. The Todos Santos coastal plain is an area of approximately 150 km² occupied by the city of Ensenada, Valle Maneadero, and Bahía de Todos Santos. This province appears to be a half graben with the Agua Blanca fault forming its southwestern border. This half graben is filled with more than 2,000 m of sedimentary strata of unknown age (Robert McEuen, 1972, oral commun.). The mesas to the north do not pass beneath these deposits but stand as high bluffs that were once wave-battered headlands.

Central Coastal Area. The broad coastal area from Puerto Tomás to Punta Falsa closely corresponds to the area underlain by Upper Cretaceous strata. Between San Quintín and Punta Canoas are four well-developed marine terraces. Near Mesa San Carlos, the terraces are capped by upper Pliocene or lower Pleistocene basalt. On Mesa San Carlos, the basalt overlies a thin, discontinuous deposit of Pliocene beach sand at an elevation of more than 600 m. Post-Pliocene erosion has removed vast quantities of Paleocene and Upper Cretaceous strata. When the Pliocene San Carlos surface was at sea level, only low hills separated the Pacific Ocean from the Gulf of California.

Llano del Berrendo. The Llano del Berrendo lies in the northeastern corner of the broad sedimentary basin, between the main Baja California Peninsula and the western ranges of the Vizcaíno Peninsula. Beal (1948) called this basin the Baja California syncline. In a few places, the basin is overlain by Pliocene or Pleistocene basalt. Coastal lagoons (Phleger and Ewing, 1962) suggest that the area is still subsiding. The lagoons form behind eolian barrier beaches and separate coastal barchan fields from sand strips and other eolian forms inland. Inman and others (1966) used radiocarbon methods to date the formation of these barrier beaches at 1,800 yr B.P.

COASTAL MOUNTAIN PROVINCES

In the northern part of the peninsula and extending into southern California, a belt of mountains lies between the coastal terraces and the interior valleys and plateaus. These ranges, containing some of the most rugged topography of the peninsula, are carved from prebatholithic volcanic and volcaniclastic rocks. This area of mature topography shows little evidence of having once been a low-lying plain; however, scattered patches of gravel, apparently derived from the east and carried to the region across the inland plateaus, imply that this region too stood close to sea level during early Tertiary time.

The western margin of the coastal mountains is marked by the easternmost limit of Late Cretaceous to Cenozoic marine deposition. In places, the deeply weathered early Tertiary surface can be seen to dip beneath the Miocene sedimentary and volcanic strata of the coastal provinces.

The eastern margin of the coastal mountain provinces is a lineament of unknown nature and origin that trends N. 30° W.; we call it the "coastal highland boundary." North of the international border, it is well marked east of Mounts San Miguel, Helix, Cowles, and Fortuna in San Diego County and may continue northeast of Black Mountain, to the west of Escondido, through Valley Center to the Elsinore trough, Riverside County. East of the Ensenada block (Fig. 38), the boundary appears to correspond to a fault with offset down to the east. Farther north, however, fault trends are oblique to the boundary, and the mountains may have resulted from a flexure as suggested in Figure 40.

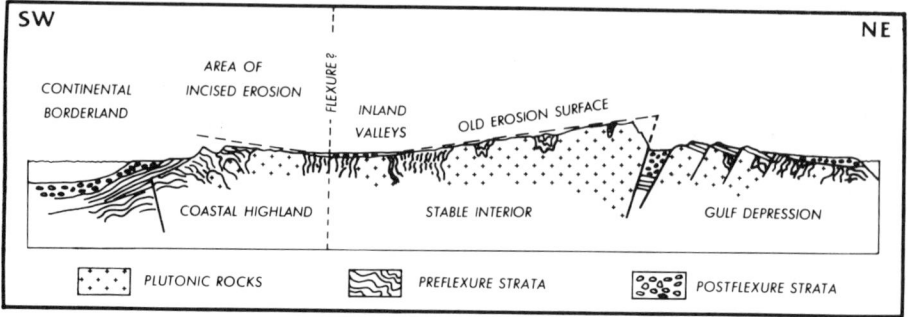

Figure 40. Diagrammatic sketch illustrating a suggested relation between the old erosion surface and the coastal highlands.

The difference in relief between the coastal mountains and the interior plateaus and highland valleys may also be due to differential erosion. The eastern margin corresponds roughly to the contact between the predominantly granitic terrane to the east and the metavolcanic rocks to the west.

La Zorra Block. The La Zorra block extends from Valle Guadalupe on the south to beyond the international border to the north. The La Zorra block descends from more than 1,000 m along its northeastern margin to 300 m where it disappears beneath the coastal deposits. Remnants of the early Tertiary surface are well preserved where it is extensively covered by conglomerate of Eocene and possibly earlier age.

Ensenada Block. The extremely rough topography of the Ensenada block extends from Valle Guadalupe south to the Agua Blanca fault. Near the eastern edge are a series of peaks: Cerro Blanco (1,355 m), located east of Valle Guadalupe; El Toro (1,405 m), located south of Valle San Rafael; and La Biznaga (1,617 m) and El León (1,700 m), both near El Alamo. The Ensenada block is dissected by the west-flowing Río Guadalupe, the Río Santa Clara, Cañon de la Chispa, and Arroyo Santo Tomás that cut gorges as much as 1,000 m deep. The Ensenada block, which stands at a higher average elevation than the La Zorra block, gives little evidence of a pre-existing early Tertiary surface. A lag deposit, the remnant of a fluvial conglomerate, is scattered over a 400-m-high plateau in the northwestern corner of the block.

Tomás Block. The Tomás block consists of the coastal mountains south of the Agua Blanca fault. South of Ensenada, the bedrock surface of the Tomás block is 1,000 m higher than the adjacent bedrock surface of the Todos Santos coastal plain. This differential elevation across the Agua Blanca fault disappears and may actually be reversed to the east. As with the La Zorra and Ensenada blocks, drainage from the northwest is through deep and narrow canyons, for example, Río Santo Tomás, Río San Vicente, Arroyo Salada, and Río San Rafael. The eastern margin is not as well marked as that of the more northern coastal mountain provinces; this clearly suggests that differential erosion is responsible for the topographic contrast between the Tomás block and the Peterson block to the east.

NORTHERN HIGHLAND PLATEAU PROVINCES

Most workers in the Peninsular Ranges have noted the striking highland plateaus that extend from the coastal mountains to the main gulf escarpment. These surfaces, mantled only by varying depths of decomposed granitic rock, form rolling highlands. They can be recognized from Orange and Riverside Counties to at least lat 29° N. North of the Agua Blanca fault (Pl. 3), these surfaces may be divided into six separate geomorphic provinces that differ in elevation, inclination, cover, and relief.

These surfaces were formed during Late Cretaceous and early Tertiary time. By mid-Eocene time, the region stood close to sea level; streams from the east carried gravel into Baja California and deposited it in a series of stream channels across the unroofed and eroded granitic rocks (Minch, 1970). The outpouring of volcanic rocks that began in mid-Tertiary time further buried the surfaces. How extensive this cover was is not known, but scattered remnants of mid-Miocene volcanic rocks are found near the international border, and the cover is nearly complete just north of Valle Trinidad. The extensive Miocene deposits in the continental borderland (Moore, 1969) suggest that the Miocene volcanic event was accompanied by an increased rate of erosion. The seismic activity of the gulf depression and the fresh scarps at the base of the main gulf escarpment suggest that the uplift is continuing.

McCain Plateau. The McCain plateau, described by Miller (1935b), extends southeastward 40 km from the Laguna Mountains of San Diego County to the La Rumorosa plateau of the Sierra Juárez and westward 30 km from the main gulf escarpment to the Tecate surface. Elevations of the plateau decline from about 1,300 m along the northeastern edge to less than 1,000 m at the southwestern edge, a slope of about 1:100. The origins of the northwestern, southeastern, and southwestern boundaries of the province are unknown. The southwest drainages are undeflected where they cross into the higher standing, more deeply dissected Tecate surface. The escarpment between the McCain and La Rumorosa plateaus probably follows a fault; the other boundaries could be due to differential erosion that is controlled either by differences in bed rock or in cover layers. Remnants of overlying Miocene sedimentary and volcanic rocks are now found only in the Jacumba area of San Diego County.

La Rumorosa Plateau. The piñon- and pine-covered La Rumorosa plateau is located southeast of the McCain plateau. It is higher than the McCain plateau but lower than the Laguna Hansen surface. The eastern area and western ridges are characteristically 1,400 to 1,500 m in elevation, whereas the intervening valleys are about 150 m lower. Many of the ridges are capped with fluvial conglomerate similar to that found farther west on the Tecate surface. About 10 km south of La Rumorosa, two small peaks rise to 1,600 m. They are capped by a thin layer of conglomerate over which is a cap of basalt. This is the only relic of volcanic strata known on the La Rumorosa plateau. Drainage across this province is predominantly southwest, but some streams have been captured by arroyos draining into the gulf depression.

Laguna Hansen Surface. Occupying the central portion of the Sierra Juárez is a pine-covered upland called the Laguna Hansen surface (Fig. 41). It is about 1,700 m high at its northeastern edge and 1,400 m at its southwestern edge, sloping about 300 m in 30 km. Local hills of 150-m relief are common throughout the province, and some canyons are as deep as 600 m.

The southwestern edge of this surface is defined by the San Miguel fault trend; its southeastern edge lies close to the main gulf escarpment. Other boundaries do not appear to be fault controlled. Differences in the original cover, in the bedrock resistance, or in the climate allowed differential denudation at some time during the erosional history. The area now drains both southwest toward the Pacific and northeast toward the Gulf of California. A remnant of the fluvial deposits that once capped the old erosion surface exists in the exotic auriferous conglomerate at Campo Nacional (loc. 7 in Fig. 20, described in Chap. 4) and at several points near the southern edge of the surface.

Tecate Surface. Between the coastal mountains to the west and the Laguna Hansen surface, La Rumorosa plateau, and McCain plateau to the east lies an area of low relief but rough topography (Fig. 14). The Tecate surface extends north of the border and merges with the rather well-defined, though deeply dissected, old erosion surface surrounding Barrett and El Capitan reservoirs (Gastil, 1961). Varied resistance of the basement rocks to erosion produced rough topography that was already present when the Eocene conglomerate was deposited. The base of the conglomerate drops from 1,200 m near the eastern edge of the province to 300 m on the western edge. The inclination of interbedded argillaceous layers suggests that at least some of this elevation drop is due to tilting of the block down to the west.

The straight western edge of the Tecate surface is marked by a chain of intermountain depressions, the most prominent of which is Valle de las Palmas. The elevation of Eocene gravel in the area of Valle de las Palmas indicates that the Tecate surface has been downwarped along its southwestern edge relative to the La Zorra block. The southwest drainage is commonly deflected where the channels cross into the coastal mountains.

Alamo Block. Situated between the San Miguel fault to the north and the Agua Blanca fault to the south, the Alamo block includes Valle San Rafael, the Alamo plain, and Valle del Rodeo. To the west, the Alamo block is bordered by the

Figure 41. North end of the Sierra Tinaja and the scarp of the Sierra Juárez. Oblique aerial view to the northwest across the north end of the Laguna Salada to the Sierra Juárez. The 1,869-m Cerro de la Parra stands above the Laguna Hansen surface (horizon). To the left (south), this old erosion surface is buried by the volcanic rocks of the Santa Catarina plateau. Downfaulted blocks capped by volcanic rocks lie in the middle distance. In the foreground, similar volcanic strata cover Tertiary sandstone and Mesozoic granitic rocks. Photograph by Dallas Clites.

coastal mountains of the Ensenada block; to the east, the bedrock surface passes under the volcanic and sedimentary strata of the Santa Catarina block.

Like the McCain plateau, the Alamo block displays extensive areas of basement rock with low relief. Most of the prominent hills are gabbro monadnocks produced by differential erosion. Some of the flatter areas, such as Valle del Rodeo, the Alamo plain, and Valle San Rafael, appear to be pediments underlain by basement rocks. North of Valle del Rodeo, isolated patches of conglomerate and volcanic rocks still cap the old erosion surface. Many of the modern drainage channels in the Alamo block have established erosional surfaces at elevations below the old surface.

The slope of the surface is primarily northwest; the elevation is 1,250 m just north of the Agua Blanca fault and 600 m at Real del Castillo to the northwest. The drainage does not follow the slope of the surface but flows southwest into Arroyo Rincón, southeast into Arroyo el Carrizo, and west into Arroyo Santo Tomás, Cañon de la Chispa, and Río Santa Clara. Only in the extreme northern end of the block is the drainage to the northwest.

Santa Catarina Plateau. The Santa Catarina plateau is a belt of volcanic-capped mesas occupying an area north of the Agua Blanca fault between the Alamo block and the main gulf escarpment. The volcanic strata rest on fluvial conglomerate and sandstone that were deposited discontinuously on a fairly regular surface of pre-Miocene basement rocks. Since deposition of the conglomerate, this surface has been uplifted and tilted and now stands at about 1,250 m. The tops of the mesas generally stand at about 1,600 m; the highest point is Cerro Félix, 1,999 m, 22 km north of San Matías Pass. Tilt is difficult to estimate because of numerous faults and variable thicknesses of volcanic rocks. North from Valle Trinidad, however, there is an almost continuous dip slope, called Mesa de Chucho Prieto, that rises northward 467 m in 12 km.

HIGHLAND VALLEY PROVINCES

Large valleys of post-Miocene origin are present in the stable provinces of the Peninsular Ranges. Some, such as San Rafael in the north end of the Alamo block, may be fault controlled. Valle Trinidad seems to be a half graben along the Agua Blanca fault. It contains at least 400 m of sedimentary fill, associated with a major negative gravity anomaly (Chap. 9). Some of these valleys, at various elevations, have pediment floors cut across several basement rock types. North of the international border those with pediment floors include Perris, Santa María, Poway, and El Cajon valleys; south of the border are Guadalupe, San Telmo, and San Rafael valleys.

Two areas of extensive pedimentation are the Llano de San Augustín (between provinces 16 and 17 in Fig. 38) and the Llano de Santa Ana (between provinces 17 and 18). Both areas have broad valley floors formed at elevations considerably below that of the early Tertiary erosion surface.

Some of the highland valleys have been occupied by lakes and are now filled with lacustrine deposits; these include the valleys at El Mármol, Rancho Cartabón east of El Arenoso, and Laguna Seca de Chapala (Arnold, 1957). Some of these lakes formed behind dams of lava that erupted during Pliocene or Pleistocene time.

SOUTHERN HIGHLAND PLATEAU PROVINCES

Although the early Tertiary erosion surface continues south of the Agua Blanca fault, it is nowhere as well preserved nor as easily determined as to the north. Deeper erosion and greater volcanic cover obscure it. The rolling, highland topography characteristic of the northern plateaus gives way to volcanic-capped buttes and mesas near lat 28° N. In general, however, elevations rise from west to east with maximum heights near the main gulf escarpment.

Drainage is generally to the west: arroyos that begin within sight of the main gulf escarpment flow westward to the Pacific, commonly through deeply incised valleys.

Peterson Block. The name "Peterson surface" was applied by Woodford and Harriss (1938) to the region around Rancho San José, northeast of San Quintín (Fig. 38). We use the name "Peterson block" to characterize the upland area of basement rock that extends from the Agua Blanca fault to the Llano de San Augustín and from the high Sierra San Pedro Mártir to the Tomás block and coastal terrace.

The Peterson block is separated from the high Sierra San Pedro Mártir by an escarpment as much as 750 m high. To the northwest, it adjoins the Tomás block, but farther south its old surface extends beneath the coastal terrace. The Peterson block is separated from the Alamo block by the Agua Blanca fault, and generally stands at a lower elevation than the surface directly north of the fault. At Valle Trinidad, there is a half graben downdropped on the southwestern side of the fault.

The old erosion surface is in general very dissected, and most of the block is covered by younger erosional surfaces produced by differential weathering and lower base levels. At several places, patches of sedimentary and volcanic strata cap remnants of the old erosion surface. The deposits are on ridges and hilltops that stand above the modern surface. The old surface descends from around 1,300 m adjacent to the high Sierra San Pedro Mártir to around 200 m just east of the coastal terraces, about 40 km distant. The apparent tilting of the peninsula across provinces 14 and 15 is illustrated in Figure 42.

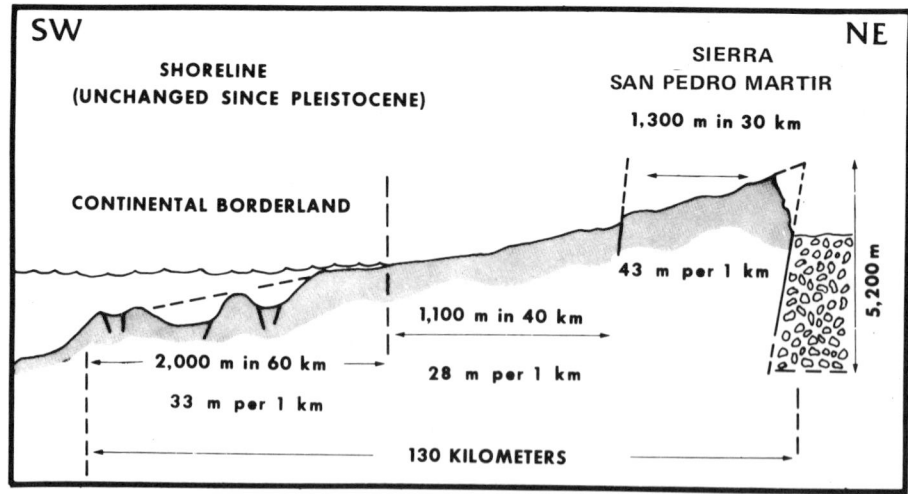

Figure 42. Diagrammatic sketch to illustrate the tilting of the peninsula along section D-D' of Figure 38 and Plate 4.

The southeastern corner of the old surface is concordant with the basement terrane on which the sedimentary and volcanic strata of the volcanic Matomí plateau (province 16 in Fig. 38) were deposited.

Sierra San Pedro Mártir. The fir- and pine-forested plateau called the Sierra San Pedro Mártir occupies the axis of the peninsula south of the Agua Blanca fault. Its northwest side is separated from the Peterson block by an impressive west-facing escarpment that rises from elevations of 1,500 m to 2,250 m or more in 1.5 km. This escarpment continues southeast for 60 km, becoming less distinct. South of lat 30°45′ N., there is no obvious topographic break, but the highland plateau still stands some 500 m above the general level of the Peterson block. The area between the Sierra San Pedro Mártir and the Peterson block has been deeply eroded by the headwaters of the Rio del Rosario and the Arroyo San Simón. Farther south, the surface of the Sierra San Pedro Mártir drops to 1,500 m and passes beneath the volcanic strata of the Matomí plateau (Fig. 43). The high surface decreases in elevation from northwest to southeast on a gradient of 1,000 m/60 km.

The surface of the Sierra San Pedro Mártir is also tilted southwestward, normal to the axis of the peninsula. At its widest point, it falls from 2,900 m at the main gulf escarpment to 1,500 m at its western escarpment, some 25 km distant (Fig. 42; Pl. 4, profile D-D′). The tilted surface apparently extended at least 5 km northeast to the 3,095-m summit of La Encantada (Picacho Diablo), the highest point in Baja California. This implied tilt of 60 m/km is considerably more than the slope of the peninsula as a whole.

The rugged gulf escarpment at the east edge of the Sierra San Pedro Mártir drops to Valle San Pedro (Valle Chico) 500 m above sea level, in a distance of a little more than 10 km. Fresh fault scarps at the foot of the escarpment imply that the uplift and tilting are continuing.

The gently rolling highland plateau is underlain by granitic rocks, with a scattered cover of alluvium. Tertiary deposits have not been found. Valleys with meandering streams and cienagas are locally developed on this surface.

Matomí Plateau. The volcanic rocks of the Matomí plateau form a nearly horizontal, almost continuous cover that extends from the main gulf escarpment westward for more than 40 km (Figs. 29 and 43). To the north, volcanic rocks bury the southern extension of the Sierra San Pedro Mártir; to the west, they bury the Peterson block. The volcanic rocks are limited to the south by the pediment-floored valleys of the Llano de San Augustín. A few northwest-trending normal faults of minor displacement (down to the east) traverse the eastern part of the plateau, but like the Santa Catarina plateau to the north, the Matomí plateau is part of the stable peninsula.

This broad, deeply dissected region of volcanic mesas is dominated by the old volcanic center of Pico Matomí, which rises 450 m above its surroundings to an elevation of 1,750 m (Fig. 43). Cerro San Juan de Dios (1,260 m) near the southern margin and Cerro Chato (1,965 m) at the northern margin also form prominent landmarks.

The southern and central portions of the plateau are at elevations between 1,000 and 1,400 m. North from Pico Matomí, however, the plateau rises about 300 m in 20 km; this continues the southeastward slope of the surface of the Sierra San Pedro Mártir noted above. The volcanic section of the Matomí plateau thickens to the northeast as the basement surface tilts (or more likely steps) down to the northeast.

The drainage is predominantly to the southwest even though the headwaters

Figure 43. Exhumed erosion surface, south end of Sierra San Pedro Mártir. Oblique aerial view to the south across the main gulf escarpment and the southern end of the old erosion surface of the Sierra San Pedro Mártir to the Matomí volcanic plateau. Pico Matomí is on the left horizon with Cerro San Juan de Dios in the distance. Photograph by John Shelton.

of Arroyo Grande rise within 1 km of the main gulf escarpment. Some of the westward-flowing tributaries actually breach small fault scarps that face the gulf in order to drain to the Pacific.

Jaraguay Block. The Jaraguay block is a large and varied province extending from the Llano de San Augustín in the north to the Llano de Santa Ana in the southeast and from the main gulf escarpment to the Pacific coast.

As in the northern areas, an erosion surface with much local relief was developed across the Jaraguay block by Miocene time. It would appear from the 1:500,000 relief map (Comisión Intersecretarial Coordinadora del Levantamiento de la Cartográfica de la República Mexicana, 1958) that the old erosion surface inclines from about 800 m near the main gulf escarpment to 400 m a few kilometers inland from the Pacific coast. If this surface was ever covered by sedimentary or volcanic deposits of Miocene age, they have been stripped away (Fig. 44). The extensive volcanic cap is almost entirely of Pliocene and Pleistocene age. In places, Pliocene erosion cut canyon floors well below the depositional surface on which the Pliocene basalt was deposited. Some Pleistocene basalt was deposited in canyons well below the Pliocene depositional surface. Today, intermontane pediments are forming 100 m or more below the general Pliocene surface.

Although most of the Jaraguay block stands above the Llano de San Augustín and Llano de Santa Ana, all of the major drainage channels rise along the main gulf escarpment and flow almost directly southwest; examples are Arroyo Santa Catarina, Arroyo Jaraguay, Arroyo Cuervito, Arroyo San José, and Arroyo de las Palomas.

San Borja Block. The San Borja block extends southeast from the Llano de Santa Ana into the territory of Baja California and constitutes the portion of the stable peninsula between the Llano del Berrendo and the gulf depression. It displays a pre-Miocene, west-sloping bedrock surface discontinuously overlain by Cenozoic sedimentary and volcanic strata. As in the Jaraguay block, much of the Miocene or earlier strata was stripped away in Pliocene and Pleistocene time, and in places, the Pliocene erosion surface has been buried by volcanic rocks of younger age. Elevations range from 2,000 m at Cerro Sandia to sea level near lat 28°30′ N.

The buried bedrock surface is very irregular. It is not known to what degree this reflects pre-Miocene or pre-Paleocene erosional or fault relief. Considerable evidence exists for faulting that postdates deposition of Miocene sedimentary strata and predates Pliocene and Pleistocene basalt. In the Sierra Santa María and Sierra San Borja, mesas extend uninterrupted from the main gulf escarpment to the Pacific coastal plain. In this 40 km, the elevations of the mesa tops drop from about 1,600 m to 500 m. The drop in elevation of the basement unconformity is considerably less, and much of the inclination in the overlying strata could be original depositional slope. Granitic rocks form an exposed surface from the volcanic mesas of the Sierra Calmallí southwest to Arroyo de San Luis. This stripped surface inclines southwestward from 900- to 300-m elev in 20 km.

Most of the province drains west or southwest to the Pacific. The long westward

Figure 44. Typical bedrock surface on the stable peninsula (Jaraguay block). Many such surfaces are apparently due to the erosion of as much as 30 m of grus from exhumed erosion surfaces originally formed in Late Cretaceous–Paleocene time. The tall tapered plant is the cirio (*Idria columnaris*) and the tall branching cactus is the cardón (*Pachycereus pringlei*).

drainage system cut gorges as much as 600 m deep during the Pliocene Epoch. In the Sierra de la Purificación, the relief is less extreme, and the drainage divide is well within the eastern boundary of the province.

The main gulf escarpment is more complicated here than anywhere else along its entire length. Between the Arroyo Calamajué and lat 29° N., the position of the escarpment is not clearly defined, and the line indicating it on the map is somewhat arbitrary. South from lat 29° to 28°30' N., the escarpment trends due south, then turns an 80° angle and trends just south of east, reaching the gulf just north of lat 28° N.

GULF OF CALIFORNIA PROVINCES

From San Gorgonio Pass in the Transverse Ranges of southern California to the La Paz embayment, near the southern end of the peninsula, the stable portions of the Peninsular Ranges are separated from the gulf depression by the main gulf escarpment (Chap. 10). Along most of its length, the mountains rise abruptly 1,000 to 3,000 m from the adjoining desert or gulf floor. In the state of Baja California, desert basins and ranges lie between the main gulf escarpment and the gulf shore.

The folding, faulting, and volcanic activity of the gulf depression have complicated its geomorphic history in comparison to that of the stable peninsula to the west. The old erosion surface, so well displayed in the stable part of the peninsula, has been rotated, faulted, and buried. Only in a few localities can it be demonstrated that this surface once continued across the bed rock of the gulf depression.

Laguna Salada. The Laguna Salada is a closed basin that extends from the international border southeast 150 km between the main gulf escarpment (Fig. 41) on the west and the Sierra de los Cucapas and Sierra del Mayor on the east. It is slightly below mean sea level at its northeastern corner and 6 m above mean sea level at its spillway south of Sierra del Mayor. High tides of the gulf occasionally flow over this divide and flood low portions of the normally dry basin.

The western part of the basin is floored by tan deposits overlying a pediment of basement rock. Along the international border, at the south end of the Sierra de los Cucapas and southwest of Pozo Cenizo, pediments extend across folded and tilted deposits of Pliocene and Pleistocene age. The east-central part of the basin and the area south of the Sierra del Mayor are a combination of playa and tidal basin.

Sierra de los Cucapas and Sierra del Mayor. These barren mountains rise abruptly from sea level to as high as 1,100 m. The northern range is sliced by N. 45° W. right-lateral faults (Barnard, 1968a, 1968b); this is the only portion of Baja California in which the topography is controlled by faults of this type. The southern end of the Sierra del Mayor terminates abruptly on a straight N. 70° W. trend. This may be only an erosional feature controlled by less-competent metamorphic rocks; however, it lies parallel to the trend of the northern Sierra Pinta across the tidal flats to the south, suggesting a structural relation.

Colorado River Delta. The major depositional surface of the modern Colorado River is a triangular basin bounded on the west by the Sierra del Mayor and the Sierra Pinta (Fig. 11). The basin opens southward into the broad, shallow northern end of the Gulf of California.

Thompson (1968) made an extensive study of the Colorado River Delta area. He divided the region into three morphologic units. The westernmost is the piedmont plain that flanks the mountains. The middle unit includes 2,000 km^2 of low-lying coastal mud flats that extend from spring-tide level to 11 to 12 m below mean sea level. The upper portions of the mud flats (that is, the high mud flats) stand about 4 m above mean sea level, the range of extreme spring tides. The side of the high mud flats toward the gulf is marked by irregular and discontinuous sets of beach ridges and sloping mud flats; this is a transition zone that slopes seaward to below the range of the spring tides, where it merges with smooth, gently sloping subtidal mud flats. Thompson's third unit includes an offshore area characterized by irregular ridges and troughs.

Sierra Pinta. The northern end of the range called the Sierra Pinta is a group of low bedrock ridges almost surrounded and partially inundated by windblown sand and modern deltaic deposits (Fig. 11). The orientation of the ridges is controlled by foliation, bedding, and a set of north-northwest–trending faults.

The central part of the range consists of variously tilted volcanic rocks of Tertiary age cut by a complex network of minor faults. In the southern part of the range, the younger volcanic strata form a generally flat lying cap across older volcanic rocks and metamorphic basement rocks. Here, the major drainage flows westward away from the adjacent gulf. This may be a relic of the pregulf drainage pattern.

Sierra Tinaja. The Sierra Tinaja is located between the main gulf escarpment and the Sierra Pinta. The northern end of this range consists of untilted volcanic strata overlying granitic basement rock (Fig. 41), simply downfaulted from the stable peninsula to the west. The remainder of the range consists of northwest-trending, west-tilted blocks of Tertiary volcanic strata, only moderately dissected by erosion (Fig. 45). The entire range, however, drains to the east; these streams are beginning to capture the drainage of the graben that separates the Sierra Tinaja from the stable Peninsula.

Cerro Borrego Block. The Cerro Borrego block, south of the Sierra Pinta and Sierra Tinaja, apparently had a structural and geomorphic history different from any of the adjacent ranges. Basement rocks stand 1,000 m above the desert floor. The andesitic to rhyolitic volcanic rocks have been removed from the higher elevations, and younger basalt flows resting on the basement rock have been steeply tilted to the west.

Sierras San Felipe and Santa Rosa. The desert ranges of Sierras San Felipe and Santa Rosa consist of high granite ridges that are mainly stripped of volcanic cover. The ridges are separated by grabens filled with hundreds of meters of first volcanic and then sedimentary strata. The province is distinguished by its northeast-trending left-lateral faults, many of which are prominent erosional features. The drainage is toward the gulf, and as in the Sierra Tinaja province, the streams are beginning to capture drainage from the basins to the west. Figure 46 shows a diagrammatic profile of the Sierra Santa Rosa north of Arroyo Matomí. Here, the granitic axis of the range apparently was re-elevated after the deposition of the older rhyolite series. A Pliocene erosion surface then formed, extending from the stable peninsula across the uplifted basement. Younger rhyolite was deposited across this surface, which was then downfaulted toward the gulf.

Valles San Felipe and San Pedro. The Valles San Felipe and San Pedro form a 100-km-long depression between the main gulf escarpment and the Sierras San Felipe and Santa Rosa province. The southwestern half of this depression is a graben; the northeastern half is mainly a pediment cut across Pliocene and Miocene strata in the Valle San Pedro and across granitic rocks in the Valle San Felipe. Despite the amount of alluvial material emanating from the eastern (that is, the

Figure 45. Oblique aerial view west-southwest across the southern end of the Sierra Tinaja to the southern Sierra Juárez on the horizon. The high volcanic Santa Catarina plateau stands at an elevation of close to 1,800 m with a peak rising to 2,000 m near the left (south) horizon. The northwest-trending main gulf escarpment crosses the upper portion of the photograph. Within the gulf depression, a succession of six north-trending, west-tilted blocks form the southern end of the Sierra Tinaja. The tilted blocks are composed of basaltic flows and andesite-rhyolite pyroclastic rocks.

main gulf) escarpment of the Sierra San Pedro Mártir, the floor of Valle San Felipe extends almost to the foot of the escarpment without a steep fan profile. This may be the result of continued subsidence along the main San Pedro Mártir fault, which would prevent the west-side fans from building outward across the valley.

Puertecitos Block. The closely faulted but nearly flat lying Pliocene rhyolite of the Puertecitos block forms a nearly continuous cover from the gulf across the main gulf escarpment onto the stable peninsula. Many of the volcanic surfaces and young fault scarps show very little erosion. The large crater of basaltic andesite on the coast south of Puertecitos is encircled to the west by faults describing a collapse caldera.

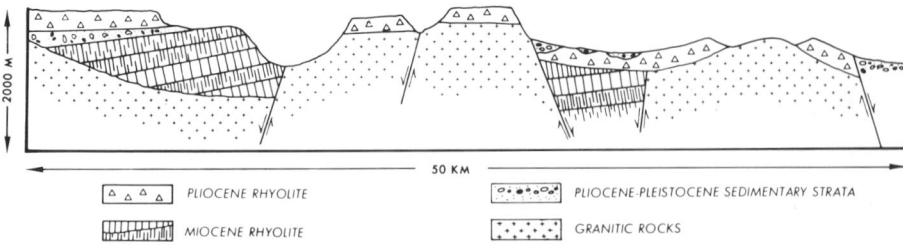

Figure 46. Diagrammatic cross section of the Sierra Santa Rosa as viewed from Arroyo Matomí.

Gonzaga Block. The Gonzaga block is primarily a depression 10 km wide and 80 km long. The main gulf escarpment bounds it on the southwest (Fig. 47), and a partially submerged ridge of granitic basement rock separates it from the Salsipuedes basin (which underlies the waters of the Ballenas Channel) on the northeast. The southernmost part of the depression is probably a thinly alluviated pediment across the basement rock, but the northern part contains Pleistocene and possibly Pliocene marine deposits and may be underlain by a deep stratigraphic basin. Fault blocks of Miocene volcanic rock west of Bahía Gonzaga and of basement rock north of Las Arastras divide the depression into three separate valleys. A fourth pediment-floored basin along the lower course of Arroyo Calamajué marks the boundary between the Gonzaga and Remedios (province 29, Fig. 38) blocks.

The rugged granitic ridges east of the depression are capped by a few flat-lying remnants of the Miocene volcanic rocks that once covered the entire area.

Remedios Block. The Remedios block lies between the Ballenas Channel and the stable peninsula, which in this area is not bordered by a continuous escarpment. The Sierra la Asamblea to the west is, however, untilted and less faulted, in contrast to the rocks of the Remedios block. The Remedios block stands highest along its western edge and steps down along northwest-trending faults into the Salsipuedes

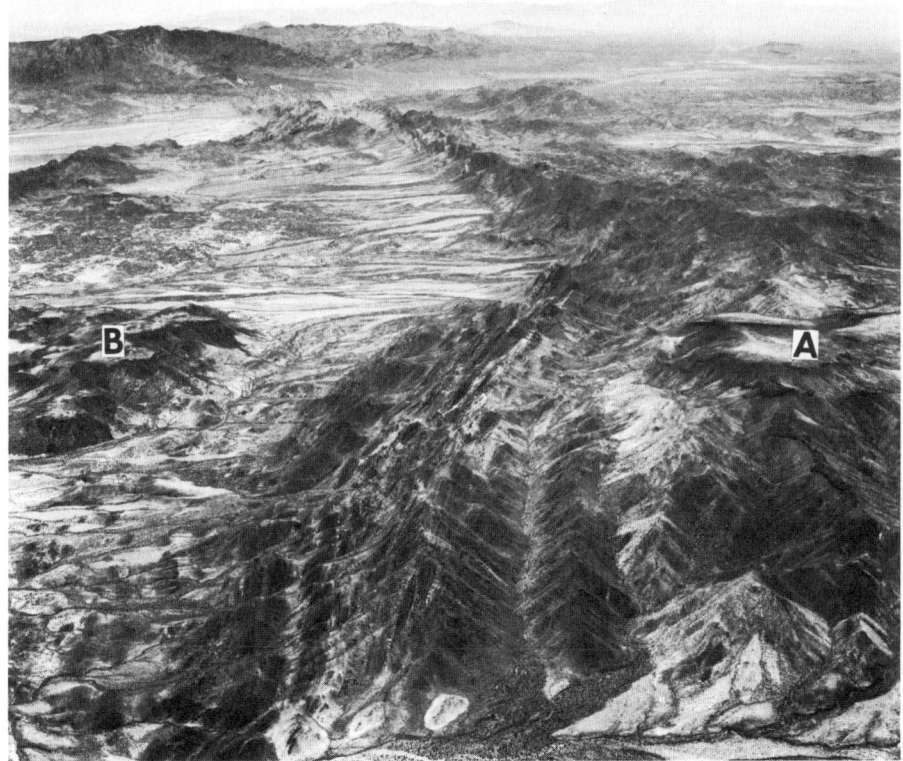

Figure 47. Intersection of Gonzaga lineament with main gulf escarpment. Oblique aerial view southeast along the main gulf escarpment toward Arroyo Calamajué. The escarpment (upper to lower center) exposes weakly metamorphosed volcaniclastic strata of unknown age. Miocene volcanic strata (A) rest on the stable portion of the peninsula about 1,000 m higher than similar strata within the gulf depression (B). At the left top across the Gonzaga lineament is the Sierra la Asamblea. Photograph by John Shelton.

basin. The Agua Amarga playas lie along the southwestern edge of the block. Fault blocks that form the gulf ranges in this area (G-G', Pl. 4) tend to be tilted toward the gulf in an antithetic pattern. Many fault blocks had been denuded of Miocene strata before the eruption of Pliocene and Pleistocene basalt (Fig. 37), whereas adjacent fault blocks retained Miocene deposits.

Los Angeles-Las Animas Block. The faults of the Los Angeles-Las Animas block trend south. The uplifted areas are generally tilted to the west and have undergone less erosional stripping than in many of the other ranges in the gulf depression. The southward trend of these structures is truncated abruptly by the main gulf escarpment where it abruptly swings S. 80° E. This nearly due east trend continues for 20 km beneath the gulf (Shepard, 1950).

El Barril Block. Located in the southeastern corner of the state of Baja California and east of the main gulf escarpment, the El Barril block is characterized by thinly alluviated pediments cut across basement rock, Tertiary volcanic rock, and (in places) Pliocene sedimentary strata. The escarpment is not as pronounced as in most of the northern provinces; south of El Barril, it essentially coincides with the shoreline of the gulf.

Near Bahía San Francisquito, a succession of marine terraces has been cut into the basement rock (Fig. 33). Directly south, Pliocene and Pleistocene marine deposits were mapped on terraces at similar elevations.

Isla Angel de la Guarda. Separated from the mainland by the Ballenas Channel (Salsipuedes basin), Isla Angel de la Guarda does not reflect the south trend of the Los Angeles-Las Animas block to the south. Isolated remnants of flat-lying younger volcanic rocks capping the interior of the island imply that the province has not undergone extensive Pleistocene tilting. In the southern part of the island, the east-flowing drainage rises near the southwestern edge; this suggests that a truncated erosional surface predated the separation of the island from the mainland (Phillips, 1966).

9
Geophysics

Until recently, few geophysical measurements had been made in the state of Baja California. The work in the Vizcaíno desert (Mina, 1957) extended 30 km north of lat 28° N. Surveys in the Imperial Valley of southern California had been extended south into the northwestern part of the state of Baja California (Kovach and Monges, 1961; Kovach and others, 1962; Biehler and others, 1964). The Consejo de Recursos Naturales no Renovables (1965) completed an airborne magnetometer survey of the region between Ensenada and San Quintín.

Extensive geophysical work was done in the continental borderland west of the peninsula (Krause, 1965; Moore, 1969) and in the Gulf of California to the east (van Andel and Shor, 1964; Phillips, 1964b).

As part of our study, a gravity survey of the state of Baja California was undertaken. The resulting Bouguer gravity map (Pl. 5) represents the work of many field crews under the supervision of one of us (Phillips), who also directed limited magnetic studies in areas of special interest.

HEAT-FLOW SURVEYS

Many measurements have been reported from heat-flow stations on the continental borderland and the adjoining Pacific Ocean floor west of Baja California (Von Herzen, 1964; Von Herzen and Maxwell, 1964). An analysis of data from 45 stations west of southern and Baja California (Von Herzen, 1964) showed that high heat flow occurs near the coast, where the average heat flow is nearly twice the normal oceanic value. Values from all of the stations in the continental borderland averaged 2.07 HFU (1 heat-flow unit = 1×10^{-6} cal/cm^{2}sec). There is a systematic decrease in heat flow in the basins off the coast of southern California, probably because of a higher sedimentation rate. Von Herzen found no correlation between topography and heat flow.

The geothermal heat flow through the floor of the Gulf of California is also significantly higher than the world-wide average of 1.5 HFU (Lee and Uyeda, 1965). In the southern part of the gulf, five stations (Von Herzen, 1963) averaged 3.42 HFU. On the basis of data from 18 heat-flow stations in the northern gulf, Henyey (1971, p. 135) concluded that:

With the exception of Wagner basin (the northernmost basin in the north Gulf) the heat flow in the northern Gulf is uniformly high with values typically in the range 2.0 to 3.5 HFU. . . .

Three locations, each near a major fault scarp, had values exceeding 4.0 HFU. The uncorrected data from one measurement in [the Salsipuedes] basin is on the order of 10 HFU. . . .

The heat flow in Wagner basin appears normal with values in the range 1.0 to 1.6 HFU.

The high regional heat flow in the Gulf of California continues north of the international border into the Imperial Valley. Combs (1971) concluded that the average heat flow in the Imperial Valley is 1.9 to 2.0 HFU. Superimposed on this broad geothermal high are local anomalies, such as the Dunes anomaly that has an observed heat flow of 40 HFU. One local anomaly in Baja California, Cerro Prieto geothermal field, is being developed as a source of power (Mooser, 1964; Mercado, 1969).

On the Pacific coast, the thermal springs south of Ensenada appear favorable for similar development. The hot springs suggest a high geothermal gradient, and the thick sedimentary fill of the Todos Santos coastal plain offers possibilities for a heat reservoir (A. L. James, App. 2; Petrick, App. 2; Strandstra, App. 2).

Roy (1963) reported a geothermal profile across the Peninsular Ranges just north of the international border. The heat-flow values were above normal and varied inversely with elevation. No heat-flow measurements south of the international border have been published, but the distribution of heat-producing elements in the Sierra San Pedro Mártir has been investigated by Smith and others (1971), who showed a rough correlation of heat-producing elements with the "felsic mineral content" of the rocks and a weak trend toward increasing heat production with elevation.

MAGNETIC SURVEYS

Krause (1965) surveyed the magnetic properties of the continental borderland off northern Baja California. He listed (p. 630) six major anomalies:

(1) An anomaly associated with the margin of the subaerial portion of the continent which is undoubtedly related both to faults trending parallel to the coast and to extensive lava extrusions, both ancient and modern, at the margin.

(2) An anomaly near the edge of the borderland north of 30°, which again must be associated with faults trending parallel to the coast and with lava extrusions at the margin of the borderland. The anomaly is not large, but it is long and continuous. It possibly continues southeast as the anomaly associated with the San Benito fault zone. (By way of contrast, no anomaly is associated with the continental margin off northern California and Oregon.)

(3) An anomaly associated with the Agua Blanca fault. . . .

(4) An anomaly associated with a very large fault zone trending southeast along the San Benito Ridge through East San Benito Island and thence through Cedros Island and into Punta Eugenia. Glaucophane schist and serpentine crop out in the fault zone at East San Benito Island; West San Benito consists of sheared volcanic rocks and grayv.ackes. This prominent anomaly extends to the northwest end of the ridge and abruptly disappears. . . .

(5) Anomalies associated with local features are related to both faults and volcanoes. The volcanoes sampled are basaltic, and the strength of the anomalies over some local hills indicates that other basaltic volcanoes are also present. Basalt occurs commonly on the peninsula.

(6) Very large anomalies in and near Bahía Sebastián Vizcaíno and in the northern borderland.

The R/V *Orca* of the Scripps Institution of Oceanography was used to obtain total magnetic-field values in the Gulf of California (Hilde, 1964; Phillips, 1964b). A proton-precession magnetometer was towed 150 m (500 ft) behind the ship. A strong correlation between topographic and magnetic anomalies was found; for example, the volcanic island of Tortuga northwest of the Guaymas basin is associated with a very large positive anomaly. At the mouth of the gulf, a series of three highs are associated with ridges or small rises that stand above the gently undulating bottom. The sharpness of the anomalies and the close correlation between these and the rises indicate that the topographic highs may be related to volcanic activity on the sea floor.

Large negative anomalies are associated with some of the submarine scarps. An anomaly of -400 γ with a gradient of 100 γ/km occurs over the northeast slope of the Guaymas basin. Another negative anomaly is associated with the north side of the Pescadero basin. If the scarps are composed of rocks that have a high magnetic susceptibility, then a negative anomaly would be expected over a south-facing cliff in the northern hemisphere. From the Farallon basin south, the deep-water areas of the gulf are associated with the positive magnetic anomalies and steep magnetic gradients. To the north, however, the basins are generally associated with small or slightly negative anomalies, and the gradients are usually lower. This effect may be attributed to the presence of a thicker layer of unconsolidated sediment masking the rock of high magnetic susceptibility. This interpretation is supported by seismic evidence.

Under the direction of the Consejo de Recursos Naturales no Renovables (1965), aerial magnetic lines were flown over the belt of prebatholithic volcaniclastic rocks between Ensenada and San Quintín as a part of a mineral exploration program. The data were presented on anomaly maps with contours at intervals of 100 γ and a scale of 1:50,000. The maps show many anomalies with large magnitudes and steep gradients. A close correlation exists between the anomalies and topographic highs, and some of the anomalies can be correlated with basalt. The high near-surface susceptibility makes it difficult to interpret the data for the effects of deeper structures.

Ground magnetometer surveys have been carried out in three restricted areas of Baja California, each in conjunction with gravity surveys. A steep magnetic gradient was found along the eastern edge of the San Quintín plain; this suggests a steep basement gradient, perhaps marking the position of the Santillán y Barrera line (Stephen Levy, 1970, oral commun.).

In the gulf depression, magnetic surveys have been used to complement gravity surveys in two of the enclosed basins at the foot of the main gulf escarpment. East of the Sierra San Pedro Mártir in the Valle San Felipe, the plutonic and metamorphic basement rocks have a relatively uniform susceptibility of about 15 $\times$ 10^6 MKS units (Slyker, 1969, App. 1). These basement rocks are intruded by and overlain by basalt and andesite. The basalt shows measured susceptibilities near 670 $\times$ 10^6; the andesite, 204 $\times$ 10^6. These contrasts made it possible to correlate the magnetic and gravity data (see below) and to make independent depth determinations. Some of the magnetic highs along the east side of the valley appear to be associated with local volcanic pipes and flows in the sedimentary section; other highs are related to dikes in the basement rock. Although the magnetic and gravity trends are roughly parallel to each other and the magnetic anomalies near the edges of the valley could be correlated with gravity anomalies, the gravity low near the center of the valley does not correspond to a magnetic low.

Farther north in the Laguna Salada, the magnetic contours show little systematic variation (Kelm, App. 1). They do not correlate with the gravity data, structure, or rock type. Even observable basaltic bodies fail to give consistent magnetic signatures.

EARTHQUAKE SEISMICITY

Baja California, together with its continental borderland to the west and the Gulf of California to the east, is one of the Earth's most seismically active areas. The determination of earthquake epicenters, however, has been hampered by the fact that, until very recently, most reporting stations lay north of Baja California at a small angle to the trend of the peninsula.

Allen and others (1960) showed the distribution of epicenters of earthquakes with magnitudes of 4.0 and greater that occurred between 1935 and 1954 in the northern part of the state of Baja California. Krause (1965) showed some additional strong events on a small-scale map of the peninsula. Sykes (1967a, 1967b) showed the larger events that occurred in the Gulf of California from 1954 to 1962.

Figure 48 shows the epicentral locations for approximately 150 earthquakes of magnitudes greater than or equal to 4.0. Most of the epicenters are in northern Baja California and in the northern part of the Gulf of California. This apparent decrease in activity southward from the international border may not reflect a decrease in tectonic activity but rather a decrease in the ability to detect small earthquakes from seismographs in the United States. However, the northern portion of the peninsula shows most of the epicenters and also contains most of the mapped faults (Pl. 3).

The epicenters in the Gulf of California cluster off the mouth of the Colorado River and form an alignment from there toward the south, leaving the area northwest of Isla Tiburón on the east side of the gulf practically free of seismic events (Fig. 48). An alignment of epicenters appears to follow the general trend of the Guaymas lineament (Fig. 55) that separates the shallow northern part of the gulf from the deeper southern part. Among the first-motion studies reported by Hodgson (1959) are those for two earthquakes that occurred at the same place along the Guaymas lineament at the south end of the Salsipuedes basin (which underlies the Ballenas Channel) on April 29, 1954; the results were identical first-motion solutions. Each solution gave a choice of two apparent fault planes: either a north strike and a left-lateral sense of motion; or a N. 45° W. strike, a 68° NE. dip, with a strike-slip component of 0.999, a dip-slip component of 0.038, and a right-lateral sense of motion. The Salsipuedes basin strikes approximately N. 45° W., and topographic and geologic evidence suggests right-lateral strike slip (Phillips, 1966).

South of the Guaymas lineament, the epicenters (Fig. 48) seem to follow the eastern edge of the continental margin. Their pattern of distribution does not form a series of en echelon zones trending either southeast or southwest across the gulf.

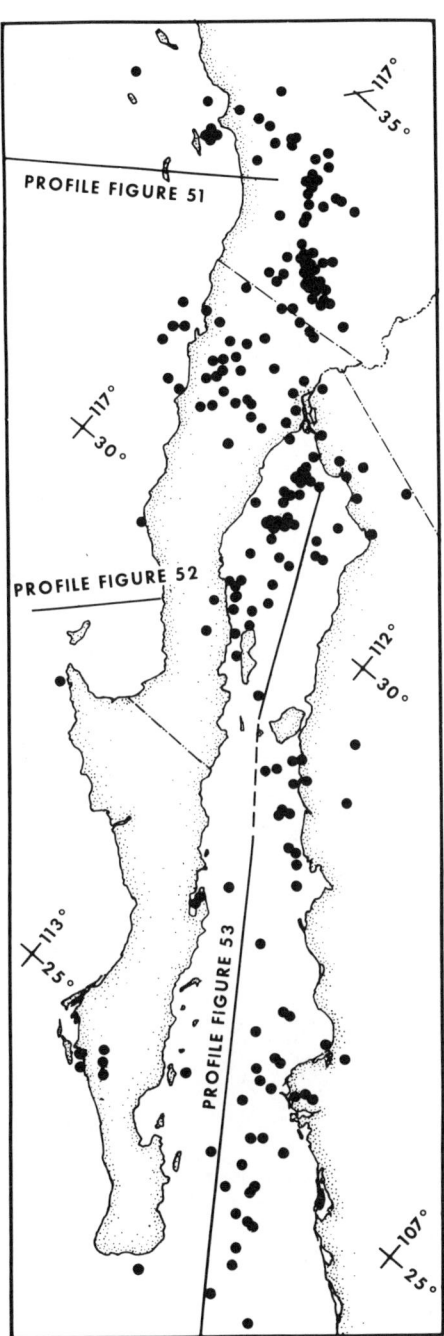

During March 1969, an unusually intense earthquake swarm occurred in the northern Gulf of California. It was studied in detail by Thatcher and Brune (1971), who used the recently installed seismograph at Rio Hardy north of the gulf and two mobile stations operated by the California Institute of Technology. Activity during this swarm was similar to that of a foreshock–main shock–aftershock sequence, but the main shock was composed of more than 70 seismic events with magnitudes between 4.0 and 5.5 that occurred during a 6-hr period. All sources were located within 5 to 10 km of each other, north of the Wagner basin and west of Consag rock (lat 31°06′ N., long 114°30′ W.; Henyey and Bischoff, 1973, Figs. 2 and 3). The hypocenters are confined to the upper crust. Focal mechanism solutions show a large component of normal faulting.

This swarm was interpreted by Thatcher and Brune (1971) as an oceanic ridge-type swarm; it thus marks an active spreading center in the northern gulf. The pattern of recent seismicity in the northern gulf suggests seismic coupling across about 200 km between adjacent inferred spreading centers.

Seismograms of earthquakes in the northern Gulf of California are clearly different from those of most northern Baja California seismic events; Thatcher (1971) interpreted this difference as a source effect. The northern Baja California sources characteristically have dimensions at least a factor of four smaller (and seismic moments an order of magnitude less) than seismic events of similar local magnitude within the gulf. Baja California events have stress drops of as much as 100 bars compared with an average of about 2 bars for the gulf earthquakes.

Figure 48. Earthquake epicenters for events of magnitude 4 or greater between 1958 and 1968 as reported by the U.S. Coast and Geodetic Survey preliminary determination of epicenters (data card service). Epicentral locations generally given to nearest 10 minutes of arc, the approximate radius of the dot on the map. Profile locations for Figures 51, 52, and 53 are shown.

Surface-wave dispersion and *P*-wave delay times have been used to study the structure of the Gulf of California and northern Baja California. Thatcher and Brune (1969) based the following conclusions on surface-wave data:

The crust in Baja California is about 30 kilometers thick and a low velocity in the upper mantle is required by the data. . . . the western edge of the Gulf is approximately oceanic with a ten kilometer crust. Crustal thickness increases to about 15 kilometers crossing the axis of the Gulf and to 20 kilometers further east in the central northern portions. . . . Sonora is similar to Baja except for a slightly higher mantle velocity and not as pronounced a low velocity channel.

P-wave delays confirm this structure (Thatcher and others, 1971) and suggest that the crust may be as much as 43 km thick beneath the Sierra Juárez.

Plate 3 shows the location of apparently young fault scarps in the state of Baja California. The scarps consist of offset alluvial surfaces and breaks in bed rock that show very little erosion. Such features are observed on the south side of Valle de las Palmas, the northeast side of Valle San Rafael, the San Miguel fault, the southern half of the San Pedro Mártir fault (Figs. 49 and 50), the alluvial flats west of Bahía San Luis Gonzaga, and several places between Bahía de las Animas and El Barril. Such scarps also appear along faults north of Punta Prieta (the Gonzaga lineament) and south of Cerro Mesquital (northeast of Guerrero Negro). Allen and others (1960) described evidence of recent breaks on the Agua Blanca fault. Faults of the Cerro Prieto trend are clearly evident for many kilometers across the Colorado River Delta.

The only well-documented surface breaks are those produced by earthquakes

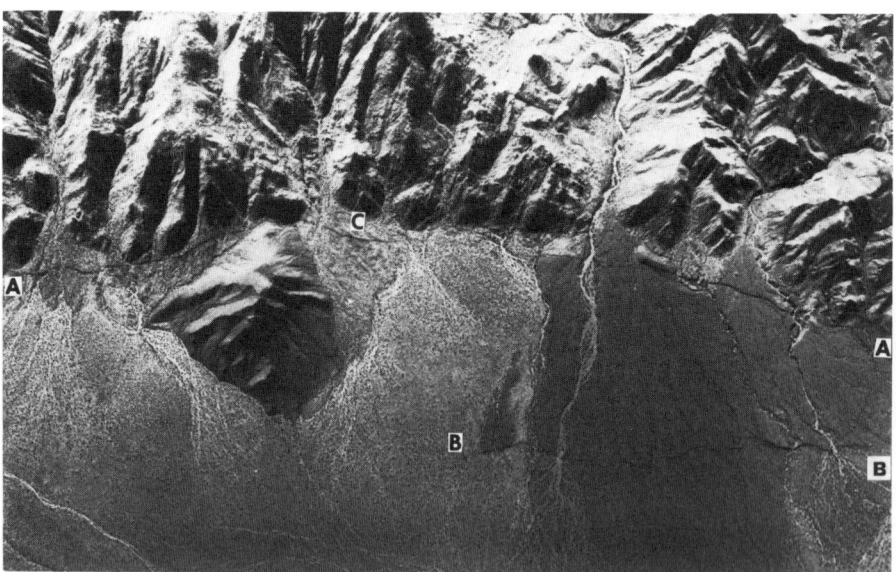

Figure 49. Aerial view of recent fault scarps along the foot of the main gulf escarpment near the south end of the Sierra San Pedro Mártir, south of Arroyo Caliente. The hill in the left center of the picture is composed of Tertiary andesite and is downfaulted about 1,500 m from the Matomí plateau to the east. It appears to rest on a fault step between the two recently active faults (A-A, B-B). The view of Figure 50, just north of the andesite hill, is marked C on the photograph.

Figure 50. Closeup of fault scarp in Figure 49. View to the southwest of a recent fault scarp along the foot of the Sierra San Pedro Mártir near the south end of Valle San Pedro. The exact location is marked on Figure 49. The maximum height of the scarp is about ten meters. The picture is from a telephoto color transparency taken by Roberta Dixon.

on the San Miguel fault in February 1956, when earthquakes of magnitudes 6.3, 6.4, and 6.8 were recorded (Shor and Roberts, 1958). Earthquakes of magnitude 6.0 or more have also been recorded near the Agua Blanca fault, near the eastern edge of the Laguna Salada, near the Cerro Prieto fault, at the south end of the Sierra del Mayor, and in the Sierra Pinta.

Several strong earthquakes were reported from Baja California during the 1800s (Wood and Heck, 1951), but their epicenters are unknown. The 1892 earthquake, whose epicenter was believed to have been located more than 100 km south of the border, was felt as far north as Visalia (between Bakersfield and Fresno, California, U.S.A.). It is believed to have had an intensity of IX to X near its epicenter. The evidence of epicenters, fresh surface breaks, and older fault lines have been combined to delineate in Plate 3 those general zones that may be seismically active.

SEISMIC SURVEYS

No explosion-induced seismic data have been reported from the state of Baja California, but extensive reflection and refraction studies have been made in the Gulf of California and the Pacific continental borderland.

Shor and Raitt (1958) constructed a profile for the continental borderland just north of the international border (Fig. 51). There is an abrupt increase in crustal thickness at the Patton Ridge followed by a gradual thickening across the continental borderland. Crustal velocities are mainly greater than 6.6 km/sec, and mantle velocities are 8.1 to 8.2 km/sec.

Figure 52 shows the profile reported by Fisher and Hess (1963) for the continental borderland near the southern end of the state. At the outer edge, under Isla San Benito, the continental borderland is similar to the section under San Clemente

Island off southern California (Fig. 51); but farther east, the crust is thinner, and the mantle velocity is lower. Crustal velocities also appear to be lower than those for the northern profile.

Seismic refraction studies of the Gulf of California have been reported by Spiess (1963) and Phillips (1964a, 1964b). Figure 53 (after Phillips, 1964b) is a longitudinal profile of the crust beneath the gulf. This profile shows a fundamental difference between the crust of the deep gulf south and west of the Guaymas lineament and the crust of the shallow northern gulf. In the deep gulf, where the submarine topography is marked by steep-sided basins, the structure of the crust seems to consist of a highly variable layer of unconsolidated sediment underlain by a layer similar to the second layer of the deep-ocean basin (Phillips, 1967) with an average thickness of 2.2 km and a seismic velocity of 5.4 km/sec. The main crustal layer has an average thickness of 3.9 km and a seismic velocity of 6.7 km/sec. Below this, at an average depth of 9.4 km is the Mohorovičić discontinuity. The upper mantle has a seismic velocity of 7.8 km/sec. The water depth in the basins decreases northward. This is primarily due to a thickening of the sedimentary strata within the basins. The general level of the base of the sedimentary strata and the depth of the Mohorovičić discontinuity do not change greatly from the mouth of the Gulf of California to the Guaymas basin. The transition zone between the middle region and the shallow northern region of the gulf is marked by a more or less abrupt thickening of the main crustal layer and the appearance of a layer with velocities that suggest semiconsolidated sedimentary rock.

The structure of the crust in those areas of the gulf where the water depth is less than 0.5 km can be summarized as follows (Phillips, 1964b): a layer of unconsolidated sediment averaging about 3.7 km thick with a seismic velocity increasing with depth to 4.15 km/sec. Below this is a layer 4.3 km thick with an average seismic velocity of 5.53 km/sec. The main crustal velocity appears

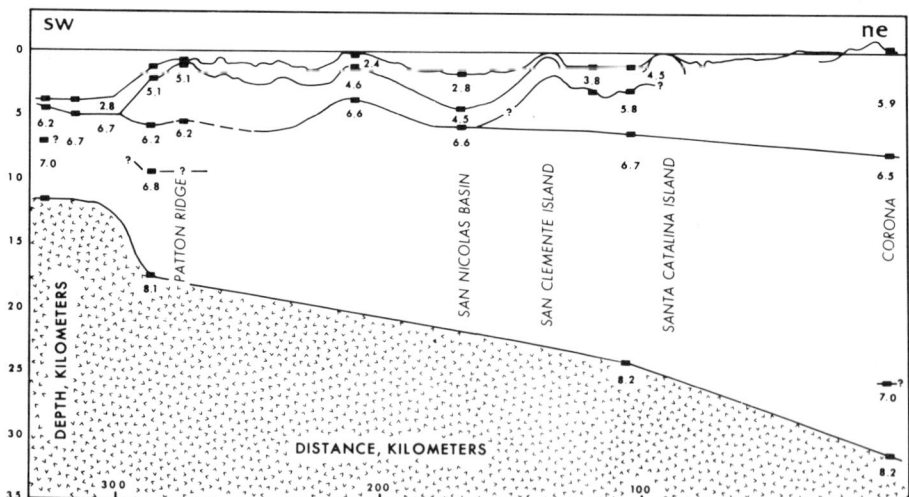

Figure 51. Seismic refraction profile across the continental borderland through San Clemente Island to Corona, California, U.S.A. (after Shor and Raitt, 1958). Black bars indicate seismic velocity control points; numbers are P wave velocities in kilometers per second. V pattern represents mantle-velocity rocks.

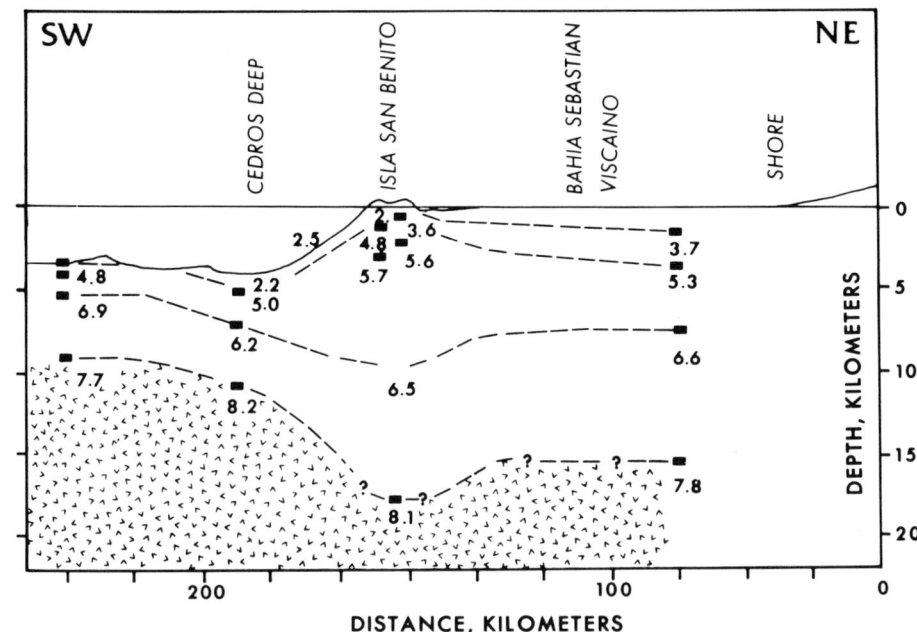

Figure 52. Seismic refraction profile across the continental borderland through Isla San Benito (after Fisher and Hess, 1963; see Fig. 51 caption for explanation).

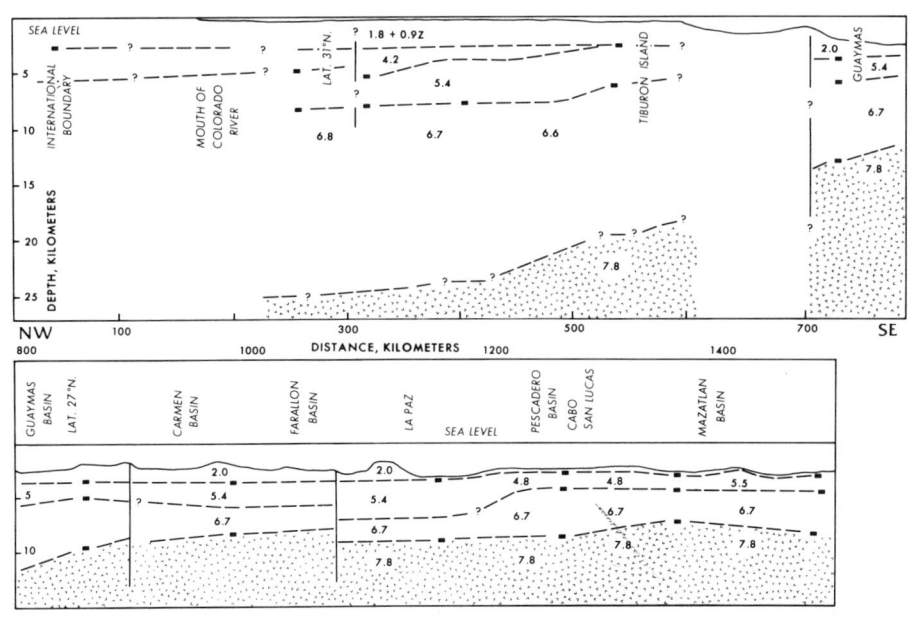

Figure 53. Seismic refraction profile along the length of the Gulf of California; vertical exaggeration ×10 (after Phillips, 1964b; see Fig. 51 caption for explanation).

to be about 6.70 km/sec. The depth to mantle in the northern gulf has not been well defined on the basis of seismic refraction but appears to be no shallower than 20 km and is probably about 25 km below the surface. On the basis of the seismic refraction work in the northern end of the gulf, the shallow structure appears to approximate a basin that is bounded by faults on the northeastern and western sides and filled with sediment. The thickest sedimentary fill (6 km) is toward the western side.

The marked thickening of the main crustal layer with a seismic velocity of 6.7 km/sec in the shallow region of the Gulf of California produces a section remarkably similar to that of the continental borderland on the Pacific side of the peninsula (Shor and Raitt, 1958).

Biehler and others (1964) reported 14 seismic profiles in the Imperial Valley, north of the international border. The profiles were intended to determine the depth of sedimentary fill beneath the Imperial Valley. Biehler and others (1964) defined as "basement" those layers with velocities in excess of 5.2 km/sec. They did not detect structures within the basement rocks. The shallow structure is similar to that of the northern gulf, with sedimentary layers at least 6 km thick.

The seismic refraction profiles within the Salsipuedes basin between the peninsula and Isla Angel de la Guarda (Phillips, 1964a, 1964b) cannot be interpreted with a high degree of precision. However, the profiles indicate that this narrow basin, which is about 1 km deep, is underlain by crust only 10 km thick and contains less than 1 km of sedimentary fill.

GRAVITY SURVEYS

A simple Bouguer gravity map for the state of Baja California (Pl. 5) is based chiefly on readings taken at almost 2,000 stations on the peninsula and on nearby islands in the gulf. Additional available data were also included in the map compilation.

Thirty-seven base stations were established (Fig. 54; App. 9) in and adjacent to the state of Baja California, using the La Coste-Romberg gravity meter no. G62. The stations were tied in with the United States gravity network through the U.S. Geological Survey gravity monument at the Los Angeles International Airport and with the network established by Chapman (1966) for southern California. Each base station in Baja California was read at least three times, and the final accepted value was derived by a least squares adjustment to the net. All base stations are considered accurate to ±0.1 mgal. The descriptions of the base stations are on file with the U.S. Army Topographic Command, Rockville, Maryland 20850.

The work of Mina (1957) was included in the southwestern corner of the state and in the territory to the south. This data was modified to match the values obtained during this survey, but in general, the two surveys were in close agreement.

The gravity contours were extended into the Pacific on the basis of information taken at sea by many authors and made available through the U.S. Department of Defense Gravity Library, Second and Arsenal Streets, St. Louis, Missouri 63118. Contours in the Gulf of California were based on the data of Harrison and Mathur (1964), recontoured in part to match the more recent information available on land and from the islands in the gulf.

During our study, gravity measurements were made with both La Coste-Romberg gravity meters lent by the U.S. Army Topographic Command and a Wordon meter

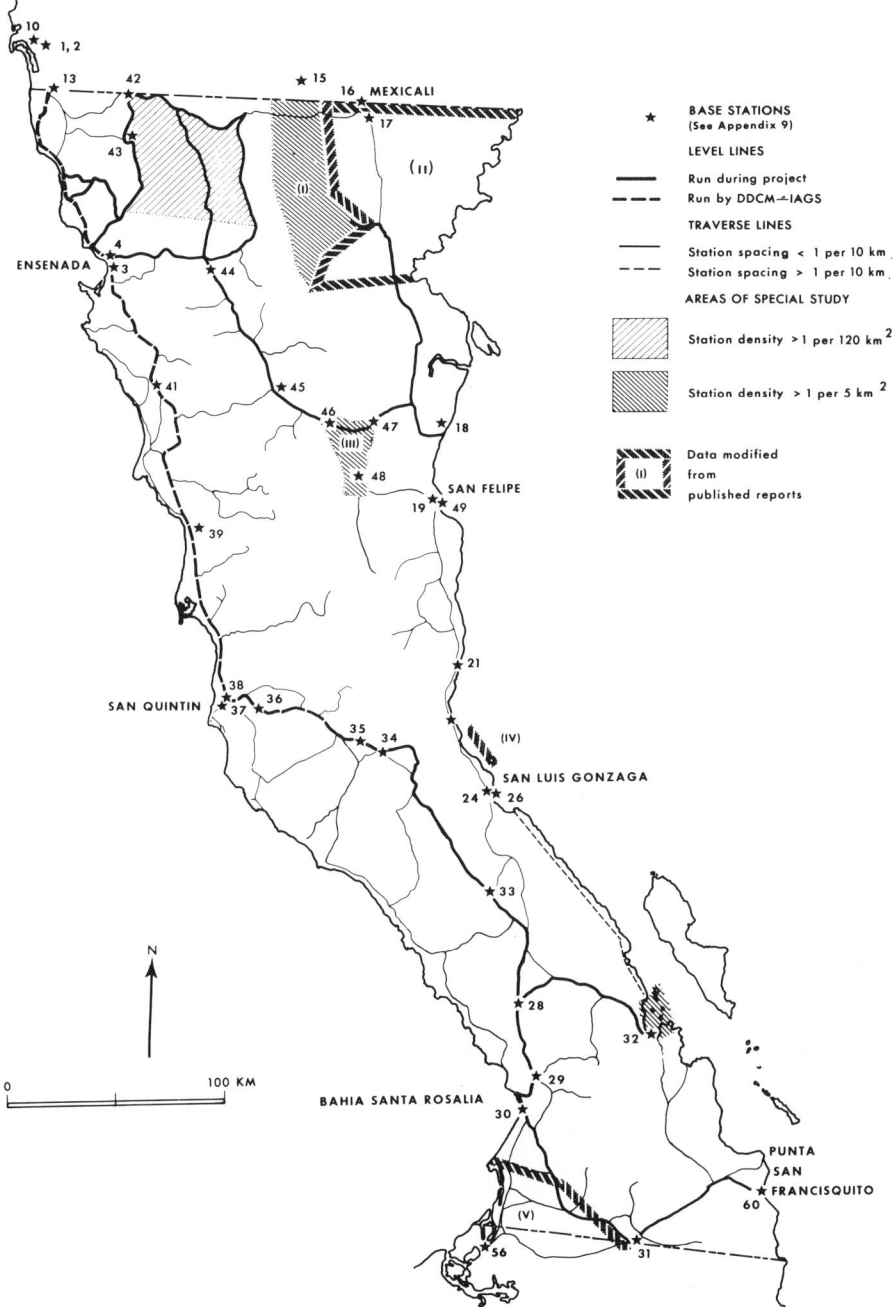

Figure 54. Gravity base stations (see App. 9), traverses, and level lines established by San Diego State University for the gravity survey of the state of Baja California. Areas with roman numerals were surveyed by I, Kelm (App. 1); II, Biehler and others (1964) and Velasco (1970); III, Slyker (App. 1); IV, Rossetter (App. 1); V, Mina (1957). Pacific coast data from the U.S. Department of Defense Gravity Library; Gulf of California data from Harrison and Mathur (1964).

owned by the California State University, San Diego. Station density varies, mainly depending on ease of access (Fig. 54). In the areas of Laguna Salada and Valle San Felipe, stations are on 1-mi (1.6-km) grids. For the northern part of the peninsula, an attempt was made to place stations on a 10-km grid. For most of the stations throughout the state, gravity measurements were made every 5 to 10 km along most roads and drivable washes. Large areas could not be covered, including much of the Sierra San Pedro Mártir and areas in the southeastern portion of the state.

Vertical control was maintained by referring to the existing bench marks set by the International Geodetic Commission and (beyond the extent of the bench marks) by leveling done specifically for this purpose, by tidal observations at stations near sea level, and by altimetry. The altimetry was controlled by reference to stations of established elevation.

Horizontal control was maintained by location on the available maps of the state of Baja California using aerial photographs and zenith camera observations. The accuracy of the elevation and location data is not as great as would be desirable for work of this kind. The absolute value of the Bouguer anomaly for any one point is probably no better than ±10 mgal, although the relative value over any given area is probably within ±2 mgal. As a result, although the Bouguer map may be warped in relation to reality, the general features are valid.

The observed gravity values were reduced to residual gravity using the international gravity formula. Free-air and Bouguer reductions were then made using a density of 2.67 g/cm^3. No topographic corrections were made because no detailed topographic information was available.

The readings of each of the individual stations and the reduced data will be placed on file with the U.S. Department of Defense Gravity Library.

No detailed analysis of the Bouguer gravity map has been completed. North of the Agua Blanca fault and west of the main batholith, the gradients are low. The discrete plutons that invade the prebatholithic volcanic rocks do not give rise to recognizable anomalies. There is a strong west-northwest trend in the gravity contours subparallel to the San Miguel fault zone impressed on the general northwest trend, parallel to the length of the peninsula.

To the east as the main batholith is encountered, the gradient becomes steeper, approaching 1 mgal/km. A gravity low of almost 100 mgal marks the crest of the Sierra Juárez. A similar low north of the international border (Kovach and others, 1962) is attributed to the mass of the low-density La Posta Quartz Diorite.

The main gulf escarpment is not marked by a strong gravity gradient. The gravity low of the Laguna Salada is not clearly separated from the Sierra Juárez anomaly. Kelm (App. 1) suggested that the Laguna Salada is a graben with 6.5 km of fill. The west boundary of the basin is not a simple fault but is composed of several steps located as much as 15 km east of the main topographic break. The northeast side, also a series of steps, is primarily defined by a nearly vertical fault located about 3 km west of the exposed fault at the foot of the Sierra de los Cucapas. The steep gradient that marks the east side of the Laguna Salada can be traced north for 40 km into Imperial County, southern California (Biehler and others, 1964). From the Laguna Salada, the gradient continues southward to the southern end of Sierra del Mayor but is not clearly expressed farther south. Gravity control is poor between the Sierra del Mayor and Valle San Felipe, but an analogous though less-extreme gradient also appears there. This steep gradient is not directly associated with any known surface fault. The gradient crosses the Laguna Salada without deflection and terminates north of the international boundary at the San Jacinto fault. Although its steepness implies that it is the result of a relatively shallow feature, Kelm suggested (App.1) that it may represent the eastern boundary of the peninsula-type crust in contrast to the denser material underlying the gulf depression.

East of the Sierra de los Cucapas, the map shows low gradients in a gravity field generally on the order of -30 mgal. A negative trough is associated with the lower course of the Colorado River; this supports the interpretation of a subsiding trough between the Sierra del Mayor and the Cerro Prieto fault. In general, the gravity data fail to express the great thickness of sedimentary fill that seismic and drill data show to exist in the central part of the basin (Biehler and others, 1964). This is possibly due to the high density of the crustal rocks that underlie the gulf depression.

The Agua Blanca fault trend is conspicuous on the gravity map. Contours are deflected where they cross the fault, and at both its east and west end, there is a marked gravity low. North of the west end of the fault, the Todos Santos coastal plain is marked by a 20-mgal low that continues out to sea. This low implies several kilometers of sedimentary rock overlying the basement rocks.

Near the east end of the Agua Blanca fault, in Valle Trinidad, there is a remarkably sharp gravity anomaly. This 40-mgal low appears to be restricted to the east end of the valley. Some of this may be due to a topographic effect because the north edge of the valley is a sharp topographic break; preliminary calculations, however, show that this can account for only about 5 mgal of the anomaly. No deep wells have been drilled within the eastern end of the valley, but irrigation wells farther west reached bed rock within 130 m.

In the western part of the peninsula, the gravity field differs markedly between the southern and northern sides of the Agua Blanca fault. Anomaly trends south of the fault are parallel to the axis of the peninsula, forming elongate, closed patterns. A prominent gravity high, as much as 35 mgal, runs continuously from the Agua Blanca fault to lat 28° N. This west-coast gravity high corresponds closely with the area underlain by prebatholithic volcanic rocks of the Alisitos Formation. Near the territory border, the trend of the anomaly apparently swings to the east and becomes more nearly parallel to the trend of the Rosarito fault (Fife, 1969).

Southwest of the west-coast gravity high, the Vizcaíno desert is marked by a deep gravity low (Mina, 1957). The Baja California syncline of Beal (1948) appears to contain as much as 15 km of sedimentary fill. This feature continues offshore along the southwestern coast of the state.

The -110-mgal low that marks the crest of the Sierra San Pedro Mártir is based on two points, one at the National Astronomical Observatory and one nearby. As in the Sierra Juárez to the north, there does not appear to be a sharp gravity gradient marking the topographic break of the main gulf escarpment. On the basis of a detailed study in the Valle San Felipe east of the escarpment (Slyker, App. 1), it was concluded that the valley represents a graben filled with 2.4 km of sediment. The fault at the western edge of the graben closely follows the foot of the escarpment, and the eastern fault is approximately 1.5 km west of the eastern edge of the Valle San Felipe.

The gravity low associated with the topographic highs of the peninsula seems to terminate against a southwest trend that marks the northern termination of the Guaymas lineament. This southwest gravity trend, which is at right angles to the peninsula, was detected in the gulf by Harrison and Mathur (1964) and is marked by a broad gravity high. Superimposed on this is a sharp gravity high associated with the small volcanic islands north of Bahía San Luis Gonzaga. These islands lie along the northeastern edge of a topographic trench that appears to be the extension of the Salsipuedes basin.

A prominent north-south gravity high crosses the gulf coast about half way between Bahía San Luis Gonzaga and Bahía de los Angeles and extends four-fifths of the way across the peninsula. This trend is parallel to and east of the Gonzaga lineament, a topographic feature with recent fault traces (Chap. 10).

South from Bahía de los Angeles is a gravity low. This trend roughly coincides with Valle de las Flores and continues south, keeping to the east of the main gulf escarpment. The low does not continue north through Bahía de los Angeles to join the low in the Salsipuedes basin. Gravity stations along the shore of Bahía de los Angeles and on the islands across its mouth restrict the contours as shown.

The low associated with the Salsipuedes basin is a well-defined feature of the gravity data of the Gulf of California (Harrison and Mathur, 1964). On the basis of seismic refraction work, topography, and bottom photographs, the basin does not appear to contain much sedimentary fill. This suggests that the basin is a young feature that has not yet reached isostatic equilibrium.

10
Structure

Lindgren (1888, 1890) was the first to delineate the major structural elements of the peninsula: the chain of partly submerged mountains along the continental borderland, the westward-inclined bedrock surface in the highlands of the northern peninsula, the impressive escarpment at its eastern margin, and the basin-and-range structure of the gulf depression (Fig. 55). Beal (Marland Oil Company of Mexico, 1924) introduced the concept of the Baja California syncline for the recurrently negative belt between Lindgren's borderland mountains and the tilted peninsula. Santillán and Barrera (1930) recognized the persistent line that has marked the position of the Pacific coast from Late Cretaceous to the present time (Fig. 55).

Shor and Roberts (1958) made the first and only detailed ground survey of a fault break in Baja California (loc. 10; Pl. 3 shows localities in Chap. 10). Allen and others (1960) described the Agua Blanca fault southeast of Ensenada, perhaps the best known fault of the peninsula (Pl. 3). Merriam (1965) investigated the "San Jacinto" fault zone in the El Gulfo area (state of Sonora) at the head of the gulf. McFall (1968) studied the Concepcion Bay fault, which follows the gulf coast south of the state of Baja California, and Barnard (1968b) described the faults of the Sierra de los Cucapas, south of Mexicali.

PREBATHOLITHIC AND BATHOLITHIC STRUCTURES

The structural framework that existed before the Mesozoic thermal event is obscure. The distribution of prebatholithic rock types (Chap. 2) suggests that the first-order structural grain existing then was subparallel to that which exists today.

Throughout most of the state, metamorphism and intrusion have modified or destroyed most pre-existing fold structures. Folds in the basement rocks generally trend northwest (locs. 29 and 42 in Fig. 4; also Figs. 12 and 47) or are semiconformable to the boundaries of plutons.

Outcrops of basement rock with folds unrelated to plutonic emplacement show a diversity of structural trends. In the northern Sierra Pinta (loc. 12; Fig. 11), McEldowney (App. 1) mapped folds that trended N. 70° W. and that were overturned to the northeast. Schroeder (App. 1) mapped east-trending folds in essentially

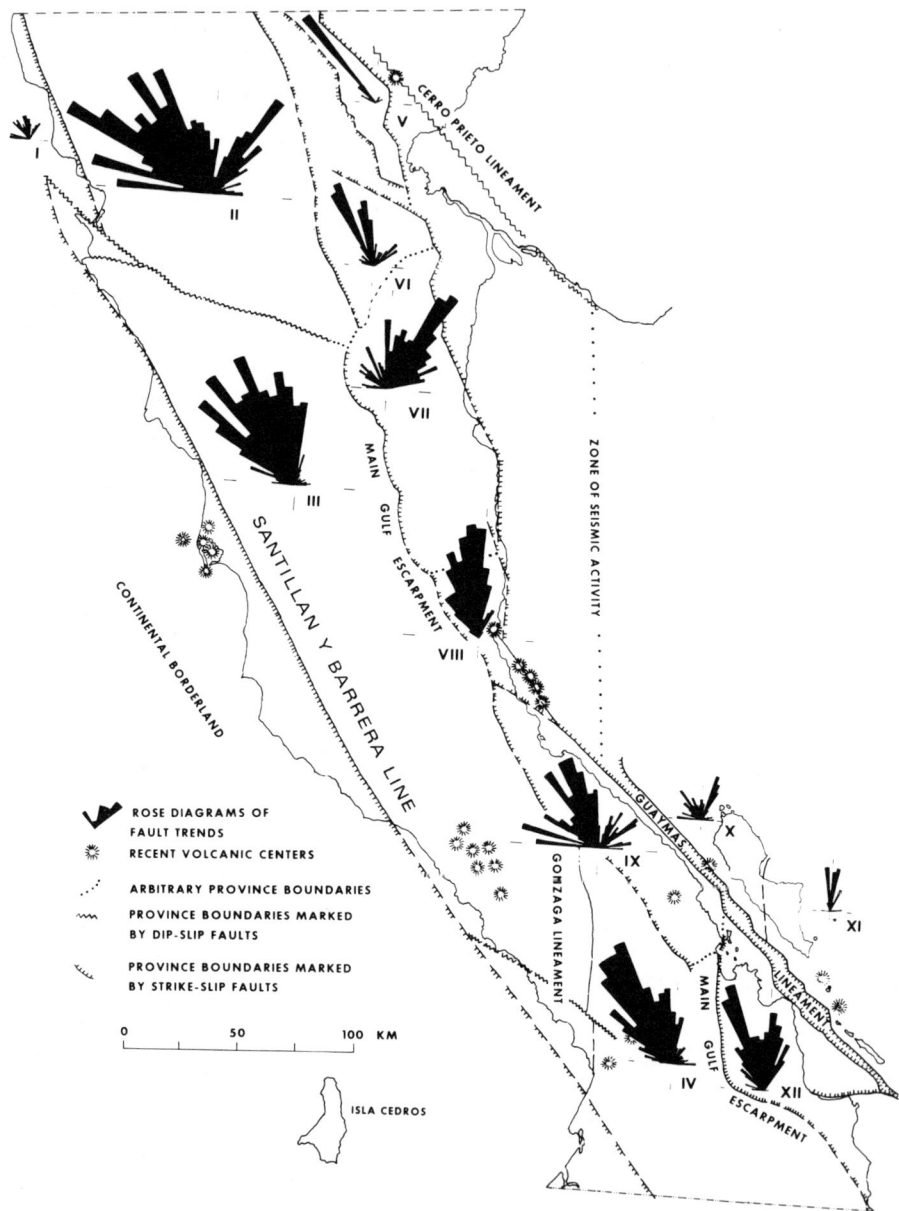

Figure 55. Structural provinces of the state of Baja California. Roman numerals correspond to provinces as follows: I, continental borderland; II, northern stable peninsula; III, central stable peninsula; IV, San Borja–Santa Gertrudis; V, Sierra de los Cucapas; VI, northern Sierra Pinta; VII, southern Sierra Pinta, Sierra San Felipe, and Sierra Santa Rosa; VIII, Puertecitos; IX, Gonzaga-Remedios coastal ranges and basins; X, Isla Angel de la Guarda; XI, Isla San Lorenzo; XII, Los Angeles–Las Animas.

unmetamorphosed volcaniclastic rocks south of Ensenada (loc. 7). South and east from El Rosario, weakly metamorphosed basement rocks strike west-northwest (loc. 20). Fife (App. 1), working north of the Rosarito fault and west of the Gonzaga lineament, mapped a prominent north-trending fold belt (loc. 24; Fig. 7).

Two areas that have been studied in detail are those of the El Pinal pluton (loc. 4; Fig. 14; Duffield, 1968) and the San José pluton (loc. 16; Birkhahn, App. 2; Murray, 1975).

UPPER CRETACEOUS STRUCTURES

The deposition of the Upper Cretaceous Rosario Formation (Chap. 4) was influenced by a structural framework that still persists. During Late Cretaceous time, the west coast of the peninsula was sharply defined, in many places at a position close to where it is today. The long and amazingly straight Santillán y Barrera line (Fig. 56; Santillán and Barrera, 1930) marks the eastern edge of the depositional trough, termed the Baja California syncline by Beal (Marland Oil Company of Mexico, 1924). The Santillán y Barrera line acted as a hinge line between the predominately emergent peninsula and the predominately submergent continental borderland.

The Baja California syncline is a depositional basin lying between the Peninsular Ranges and the positive ridges of the continental borderland. It can be traced from the Vizcaíno desert in the territory of Baja California northward along the eastern edge of the continental borderland to the Los Angeles basin of southern California. Those portions of the peninsula lying west of the Santillán y Barrera line belong to this basin. Mina (1957) reported a thickness of 6,000 m of sedimentary rock at Pozo Iray no. 2 in the northern Magdalena Plain, territory of Baja California; Kilmer (1963) reported 3,000 m of Upper Cretaceous strata at El Rosario; and Elliott (1970) reported 1,800 m of marine strata beneath the Coronado Strand just north of the international border.

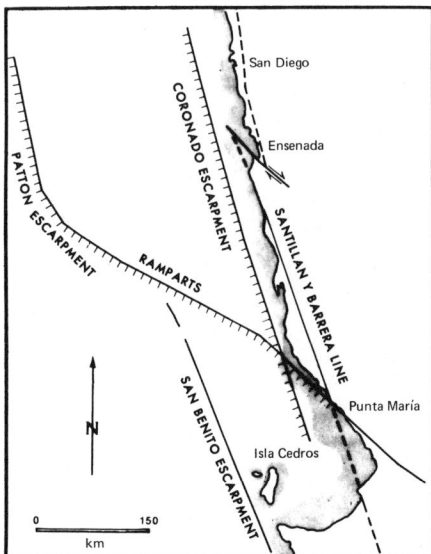

Figure 56. Major structural elements of the continental borderland west of southern California and the state of Baja California.

Acosta (1970) presented evidence that the Rosario basin near Bahía Santo Tomás (loc. 8) was an active graben during deposition; this suggests that the Agua Blanca fault system was active during Late Cretaceous time.

MAJOR STRUCTURAL LINEAMENTS

The principal northwest-trending lineaments are the Santillán y Barrera line on the west side and the main gulf escarpment on the east side of the peninsula. Transpeninsular lineaments are not as well defined. The principal ones are the Agua Blanca fault system, which strikes west-northwest, and the Gonzaga lineament, which strikes almost north.

Santillán y Barrera Line. Since Late Cretaceous time, the eastern limit of major marine transgression across the peninsula has been a line trending northwest close to the present shore (Chap. 4) This line can be followed from the Magdalena Plain on the south, north into southern California (Fig. 56). Marine deposition east of this line during Late Cretaceous and Cenozoic time is almost unknown, although near Mesa San Carlos (loc. 24 in Fig. 20) Eocene and Paleocene strata occur almost 20 km inland from the eastern margin of Upper Cretaceous marine deposits.

The Santillán y Barrera line has been a depositional hinge line, defining the eastern boundary of Beal's Baja California syncline (Marland Oil Company of Mexico, 1924) since middle Cretaceous time. A fault cannot be demonstrated anywhere along it, but the straight, boldly cliffed coast of Baja California both today and in Late Cretaceous time strongly suggests a fault-line structure. South of the Agua Blanca fault, the line is closely paralleled for much of its length by a gravity high (Chap. 9), but this geophysical anomaly diverges from the Santillán y Barrera line south of lat 29° N.

Main Gulf Escarpment. The main gulf escarpment, recognized by Gabb (1882) and Lindgren (1888), extends from Mount San Jacinto in southern California down the length of Baja California (Figs. 12, 41, 42, 45, and 47). The escarpment has been produced by movement along a network of high-angle faults. These faults can be seen in In-Ko-Pah Gorge north of the international border (north of loc. 2), on the La Rumorosa grade just south of the border (loc. 2), and elsewhere along the western edge of the Laguna Salada (Kelm, App. 1). Many of these faults dip steeply to the west and have fault gouge 2 or 3 m thick. In In-Ko-Pah Gorge just north of the international boundary, most of the fault displacement is down to the west, producing a partly antithetic fault set. The remarkable alignment of the western edge of the Laguna Salada with the medial fault of the Valle San Felipe (Fig. 57) must be inherited from an older common structure, but continuity between the modern faults has not been demonstrated. Additional descriptions of the faults along the main gulf escarpment are below in the discussion of the gulf depression structural provinces.

Agua Blanca-Santo Tomás Fault System. The existence of the Agua Blanca–Santo Tomás fault system was suggested by Böse and Wittich (1913) and by Beal (1948). It was described in detail by Allen and others (1960). At its eastern limit, it forms the northern escarpment of Valle Trinidad. From there, it trends N. 80° W. to Rancho Zacatón (loc. B in Fig. 58) where it swings to a N. 50° W. trend. In Valle Santo Tomás (loc. A in Fig. 58), the fault splits. The more northerly branch, the Agua Blanca fault proper, continues N. 50° W. to form the north side of

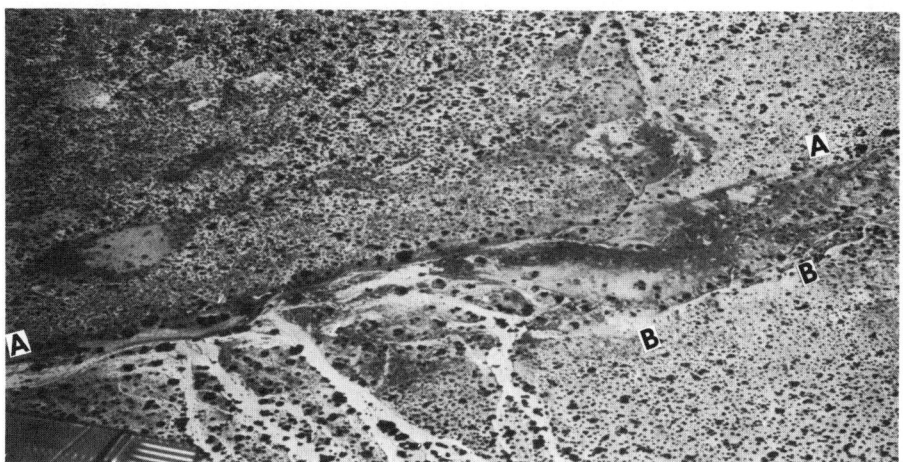

Figure 57. Medial fault in the Valle San Pedro. Oblique aerial view west across the center of Valle San Pedro (Valle Chico). The abrupt deflection of the drainages against this median depression suggests that a fault runs down the center of the valley. The lack of evidence for major vertical or horizontal movement suggests that the valley is widening by dilation along this line. Two possible fault traces (A-A and B-B) may define a small graben. Photograph by Dallas Clites.

Punta Banda, just south of Ensenada. The southerly branch, the Santo Tomás fault, follows Valle Santo Tomás N. 80° W. to the Pacific.

In the offshore area, the Santo Tomás fault curves in a gentle arc to S. 60° W. and can be traced about 20 km across the continental borderland, where it "stands as a geophysically detectable primary structural feature of the southern area" (Moore, 1969, p. 39).

Krause (1965, p. 630) noted that the Santo Tomás fault "marks the boundary between the northern and southern continental borderland. The regional depth changes profoundly across the fault. The bathymetry is dropped about 450 m to the south and appears to be offset left-laterally for 15 km."

There appears to be little evidence of recent movement on the Santo Tomás

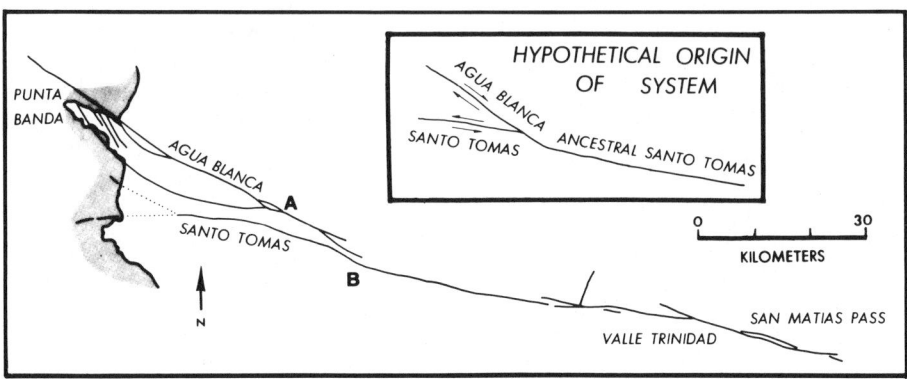

Figure 58. Agua Blanca-Santo Tomás fault system: A, junction of Valle Santo Tomás and the Agua Blanca fault; B, a point near Rancho Zacatán where the Agua Blanca fault bends toward the east.

fault. "Truncated drainage patterns associated with the submarine canyon off Punta Banda suggest a more recent displacement on the northwest-trending San Clemente (San Isidro) fault, which cuts across the older Santo Tomás fault" (Moore, 1969, p. 39).

On land, Allen and others (1960, p. 467-479) concluded that the Santo Tomás fault has a strong component of "right-handed strike-slip" together with a considerable component of vertical displacement. Work by Acosta (1970) yielded no evidence to either verify or contradict this conclusion.

Movement along the Santo Tomás fault was probably initiated early. Acosta (App. 1) wrote (p. 65) of the Upper Cretaceous deposits between the Agua Blanca and Santo Tomás faults (loc. 8):

The long, straight contacts between rocks of the Rosario and Alisitos Formations suggests that the shape of the Bahía Soledad Embayment is fault controlled. This would explain the sharp, high-angle contacts between the two formations. The faulting probably occurred prior to and contemporaneous with the deposition of the Upper Cretaceous sediments, with possibly some post-Cretaceous movement also involved.

The Agua Blanca fault can be traced offshore where it swings to a strike of N. 45° W. and merges with the San Clemente fault (Moore, 1969). The combined Agua Blanca-San Clemente fault can be traced for 260 km northwest from Punta Banda to San Clemente Island off California, U.S.A., where it forms the eastern escarpment of the island. Several large earthquakes have been associated with this fault.

Two fundamental geologic features are truncated by the Agua Blanca fault. The first of these is the belt of fossiliferous middle Cretaceous volcanic and volcaniclastic rocks (Alisitos Formation). From Valle Santo Tomás (loc. 8) 325 km south to Rancho Agua Refugio (lat 28°40′ N.), much if not most of the prebatholithic volcanic sequence is of middle Cretaceous age; north of the Agua Blanca fault, however, no prebatholithic fossils of Cretaceous age have been found. The second feature is the west coast gravity high (Chap. 9; Pl. 5), which extends from the Santo Tomás fault south at least to lat 28° N. This high ends at the Agua Blanca fault and does not reappear to the north. The origin of the gravity anomaly (Chap. 9) is unknown. The termination of these two features could be explained either by large left-lateral displacement along the ancestral Santo Tomás fault or by major uplift and erosion of the area to the north of the Agua Blanca fault. The fact that the Alisitos Formation is present between the Agua Blanca and Santo Tomás faults detracts from the first alternative, whereas the lack of an obvious change in metamorphic grade across the fault detracts from the second alternative.

Right-lateral offset along the Agua Blanca fault is suggested by the geometry of its junction with the Santo Tomás fault. If the early movement on the Agua Blanca-Santo Tomás fault system was confined to the Santo Tomás fault (which was then continuous with the eastern portion of the ancestral Santo Tomás fault and had the same N. 80° W. trend), then the recently active western Agua Blanca fault may represent a new fault that has "captured" the movement of the previous western (Santo Tomás) stretch of the system. The apparent offset along the Agua Blanca-Santo Tomás fault trend from Rancho Zacatón (loc. B in Fig. 58) to the Valle Santo Tomás (loc. A in Fig. 58) is about 20 km.

The Agua Blanca-Santo Tomás fault system cannot be traced east through San Matías Pass (east of loc. 14) into Valle San Felipe (Estavillo and Rogers, 1970; Rogers, App. 2; Slyker, App. 1).

Gonzaga Lineament. The Gonzaga lineament (Fife, App. 1) is a prominent topographic depression that runs from the Rosarito fault in the southern part of the state north to the gulf coast just south of Bahía San Luis Gonzaga (Fig. 47). The southern portion of the feature follows Arroyo León and the Llano de Santa Ana; the northern portion follows Arroyo Calamajué.

The nature and origin of the Gonzaga lineament are not clear. A number of fault breaks occur in the alluvium between Punta Prieta and Cerrito Blanco (loc. 29); this suggests a zone of active faulting. No major offsets across the lineament have been recognized.

Other north trends are the Valle de las Flores south of Bahía de los Angeles, the Valle de las Animas (loc. 27), the gulf coast between Puertecitos and San Felipe, and the zone of earthquake epicenters in the northern gulf.

Rise-Rise Transform System. Menard and Atwater (1968), Moore and Buffington (1968), Atwater (1970), and others have hypothesized a system of short northeast-trending spreading centers and rise-rise transforms extending the length of the Gulf of California. Three portions of this system are recognized in Figure 55. To the south, the Guaymas lineament extends from the northern edge of the Guaymas basin northwest past Islas San Lorenzo and Angel de la Guarda to the small islands north of Bahía San Luis Gonzaga. This lineament is expressed by enclosed basins, such as the Salsipuedes basin; ridges, such as that connecting the small islands southeast of Isla Angel de la Guarda; recent volcanism, such as that on Islas San Luis (La Encantada), Raza, and Tortuga (Chap. 6); and seismic activity (Chap. 9). There is also evidence for major right-lateral strike slip along the lineament (Phillips, 1966).

We show a north-trending belt of seismic activity through the northern part of the gulf. According to Henyey and Bischoff (1973), this corresponds to one or more linear crustal spreading centers (not called "rises" here because the areas are actually depressions lying between the Guaymas lineament and the "Cerro Prieto lineament," which is the name we apply to the transform fault connecting the spreading centers of the northern gulf with the spreading center beneath the Imperial Valley. This lineament is marked by the Cerro Prieto volcanic and geothermal center and by the faults extending from Cerro Prieto southeast to the El Gulfo area (state of Sonora) just east of the mouth of the Colorado River (Merriam, 1965).

STRUCTURAL PROVINCES

The state of Baja California was divided into 12 structural provinces distinguished on the basis of structural history and tectonic pattern (Fig. 55). The continental borderland is one large structural province. The other 11 are part of either the stable peninsula or the Gulf of California depression.

Continental Borderland

The intricate submarine topography and the structural implications of the continental borderland have been discussed by Shepard and Emery (1941), Krause (1965), and Moore (1969). In the continental borderland, there is (Moore, 1969, p. 1).

. . . a large central region of preorogenic sedimentary rock, largely Miocene in age, flanked on either side by topographic and structural highs of volcanic and basement rocks. The extensive areal continuity and relative simplicity of internal structure within these rocks suggest that their origin was by deposition on an open continental terrace or in a simple broad basin rather than in isolated block-faulted basins.

The major northwest-trending fault zones of the region are believed to have formed in middle Miocene [time] with compression as a result of wedging of the contained block causing folding in the central region. . . . Reactivation . . . in Pliocene and Pleistocene time has formed new basin and feeder-canyon systems in the inner borderland.

Krause (1965) described the seaward extensions of the Agua Blanca and Santo Tomás faults. The Rosario fault (Fife, 1969) is a major left-lateral fault striking northwest across the present shoreline in the southern portion of the state of Baja California. The seaward extension of the fault is not clearly defined by work in the borderland (Moore, 1969), but there is a topographic alignment of the north edge of the South San Quintín and Soledad basins and the Rampart Ridge. This suggests that the fault continues seaward.

The magnetic intensity map (Krause, 1965, his Pl. 2) illustrates how the offshore structure relates to the Santillán y Barrera line rather than to the present shoreline. Between Socorro and Punta Canoas, the modern coastline extends as much as 16 km west of the line (Fig. 59). Magnetic contours clearly parallel the Santillán y Barrera line and are undeflected by the modern coastline. The uplift of the coastal bulge between San Quintín and Punta Canoas must have occurred during late Pleistocene time. The features causing the magnetic pattern on the shelf must be steeply inclined and of early or pre-Pleistocene origin.

The Pacific coastal shelf, lying southwest of the Santillán y Barrera line, is an emergent portion of the continental borderland. The structure of the shelf is relatively simple, with few faults of more than minor displacement. Pliocene terraces have been warped (Orme, 1971) into a broad wavelike pattern with an amplitude of 100 to 300 m and a wave length of about 150 km. These terraces stand at their highest elevations near Punta Banda, El Rosario, and Mesa San Carlos.

Stable Peninsula

Northern Stable Peninsula. The northern stable peninsula structural province lies east of the Santillán y Barrera line and west of the main gulf escarpment; it extends from the Agua Blanca fault north to the Elsinore fault in Riverside and San Diego Counties. The Elsinore fault is the most southerly fault of the San Andreas system. Although the northern part of this province is seismically inactive, there are many faults and apparent faults that may belong to a conjugate fault system. The southern half of the province is seismically active along the San Miguel-Tijuana fault trend (Pl. 3; Shor and Roberts, 1958; Wiegand, 1970). This trend of active or recently active faults begins northeast of Valle Trinidad in the Catarina volcanic plateau (loc. 10). In this area, the San Miguel fault broke along a 20-km stretch on February 9 and 14, 1956. Shor and Roberts (1958) made a detailed map of the actual surface breaks. The fault separations were right lateral and up to the northeast. The strands of most recent rupture do not appear on pre-1956 aerial photographs. The ruptures consist of small en echelon breaks invariably stepping to the left. Northwestward from the area of the 1956 break, the fault system can be followed almost continuously to the area northeast of Valle San Rafael where offsets of streams and large dacite dikes are clearly right lateral. Raymond Elliot reported (1968, oral commun.)

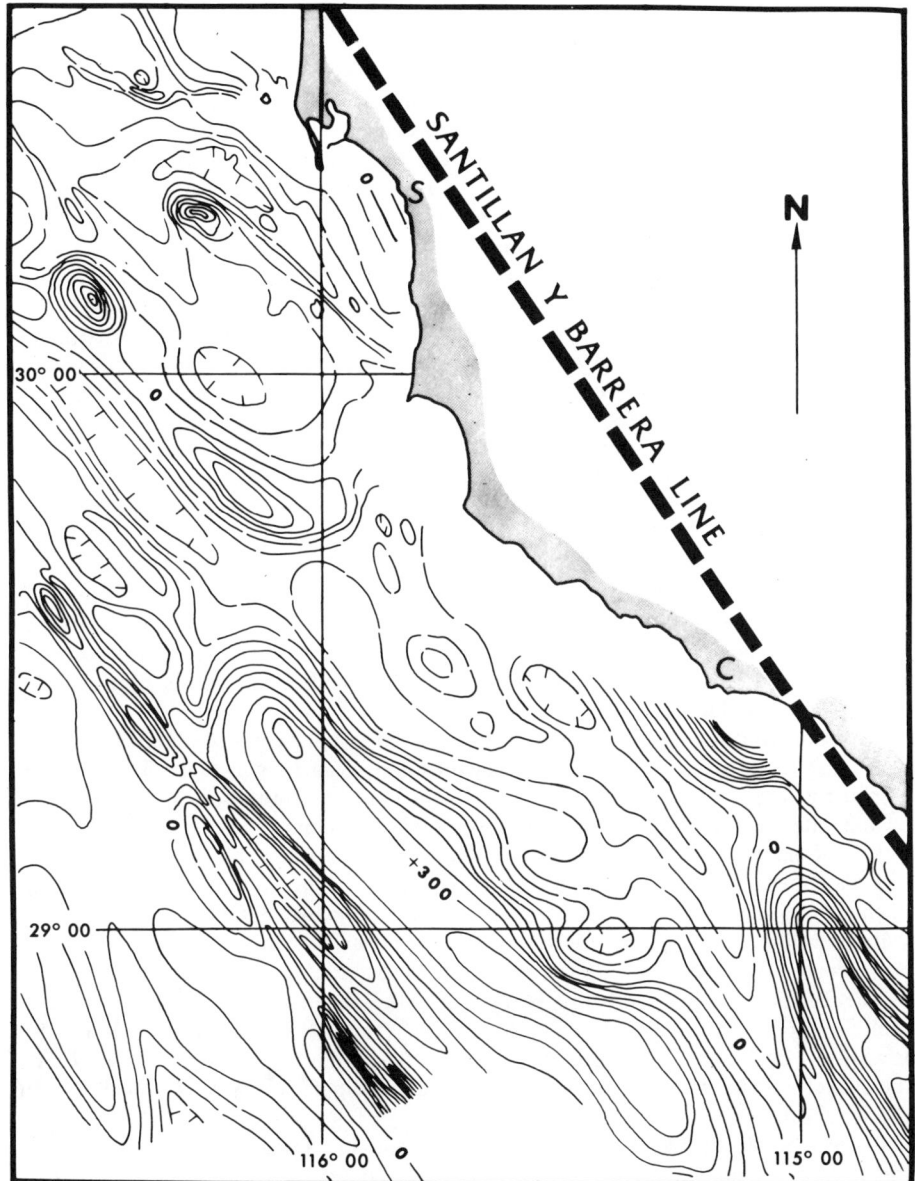

Figure 59. Relation of offshore magnetic anomalies (Krause, 1965) to the Santillán y Barrera line between Socorro (S) and Punta Canoas (C).

that left-lateral, northeast-trending faults are also present in the area. Farther north, the zone is marked by a sequence of en echelon faults stepping to the right. In the Valle Seco area (southwest of loc. 4), detailed mapping of large dikes again shows right-lateral offset.

The Vallecitos fault (Bell, App. 2; Pl. 3) extends 50 km northwest from the Sierra Juárez to the western end of Valle de las Palmas, east of Rosarito Beach.

Although the main trace of the fault is marked by erosion, there is no clear evidence that the fault moved during Cenozoic time.

About 5 km north of the Vallecitos fault, Bell mapped the Calabazas fault (loc. 32). This fault extends from the El Pinal pluton into the Valle de las Palmas area, where there is evidence of recent, possibly historic movement. There are breaks in the uplifted alluvial deposits south of the valley, small sags and springs, and at one place an uneroded set of small north-trending scarps and trenches in the bed rock. Our reconnaissance mapping did not detect a fault connecting Valle de las Palmas with the Tijuana River valley. The hypothesis that a fault underlies the Tijuana River is based on stratigraphic and structural changes across the valley. Additional evidence for faulting has been summarized by Wiegand (1970).

The fault map (Pl. 3) shows a conjugate set of northeast-trending faults throughout the province, but few of these are well established. Those mapped southeast of Tijuana by Flynn (1970) are dip-slip faults; most movements are post-Pliocene and earlier than late Pleistocene, but some are post-Eocene and pre-Miocene. The set of faults extending southwest from the Vallecitos fault to Ensenada shows a few drainage offsets that suggest left-lateral movement. Similar northeast-trending structures cross both the San Paulito pluton (loc. 5) and the El Pinal pluton (loc. 4; Fig. 14). W. A. Duffield (1967, written commun.) reported that those in the El Pinal pluton are only joints. Some of those in the San Paulito pluton, however, appear to show small offsets on the margin of the pluton. The northwest-trending faults just north of El Alamo (loc. 34) show up prominently on photographs from orbiting satellites and in regional mapping but are hard to locate on the ground. Lowman (1972, p. 34–37) illustrated similar west-trending structures north of the international border.

West and southwest of El Rodeo, there are three northeast-trending faults (loc. 9). The fault farthest west terminates against the Agua Blanca fault. These faults displace a large set of pegmatite dikes. From photointerpretation, it appears that accumulated left-lateral offset on the three faults amounts to about 1 km.

A fault can be postulated on geomorphic grounds between the coastal mountains and the interior plateaus of the stable peninsula (Chap. 8). Such a fault would begin at the Agua Blanca fault zone and trend northwest, parallel to the Santillán y Barrera line. The topographic break that suggests this fault can be traced into San Diego County at least as far north as Escondido, and possibly as far north as the Elsinore fault. The only place along the postulated fault zone where a fault is apparent is west of Valle San Rafael (loc. 6), where it was noted by Lindgren (1889, p. 6–7). Here, in the deflected valley of the Guadalupe River, the offset of old stream terraces suggests a dip-slip movement of 400 m up on the west side.

Central Stable Peninsula. Throughout the Cenozoic Era, the central portion of the northern peninsula has been the most stable area of Baja California. This structural province extends from the Agua Blanca fault on the north to the Gonzaga lineament and the Rosarito fault on the south and from the Santillán y Barrera line on the west to the main gulf escarpment on the east (Fig. 55). Cenozoic faults are uncommon throughout this area. The north-trending faults in the southern part of the province parallel major folds in prebatholithic rocks and may be of Mesozoic age.

San Borja–Santa Gertrudis. South of the Rosarito fault (Fife, 1969) and east of the Gonzaga lineament is an ill-defined structural province that combines the stable, untilted nature of the central stable peninsula, the northwest fault trends of the northern stable peninsula, and the eastward down-stepping fault blocks of the gulf structural provinces. The northern part of the San Borja–Santa Gertrudis structural province appears to be essentially devoid of faults, whereas the eastern part is cut by many high-angle faults of small displacement.

The Rosarito fault is of pre-Paleocene age with small Pleistocene adjustments. Fife (1969) traced it from the Pacific coast near Punta María (loc. 26) into the terrane of amphibolite-facies metamorphic rocks. Farther southeast, there is no suggestion of the fault for about 40 km; then a fault appears exactly on strike in the central part of the peninsula (loc. 31), becoming more nearly east-west farther to the east. Left-lateral displacement of 25 km on the Rosarito fault is suggested by the apparent offset of a concentrically zoned gabbroic pluton (Fife, App. 1). Rusnak and others (1964) postulated the "Santa Rosalía fault" crossing the peninsula through the southern edge of the province (locs. 30 to 35). This was supposed to be a transpeninsula right-lateral fault with late Tertiary activity.

Gulf of California Depression

The structures east of the main gulf escarpment are part of the Gulf of California rift system (Larson and others, 1968; Moore and Buffington, 1968). This rift system extends from the East Pacific Rise southwest of the gulf through the gulf depression, terminating to the north against the Banning–Mission Creek (San Andreas) fault. Larson and others (1968), Atwater (1971), and many others considered the gulf rift system to be related to the spreading of the East Pacific Rise and Gorda Ridge, with the San Andreas and related faults acting as rise-rise transforms. In the southern half of the gulf, the rift system is expressed by a series of basins floored by thinly mantled oceanic crust (Phillips, 1964a). These basins are believed to be small spreading centers along which the gulf has widened as new oceanic crust was added. The individual spreading centers are believed to be joined by transform faults. The strike of these faults would be northwest; and the sense of slip, right lateral. They would not be expected to extend beyond the portion of the gulf that has widened by sea-floor spreading.

In the northern half of the gulf and in the portions of the gulf depression that are subaerial, the rift basins have filled with the sediment of the Colorado River as they formed. East of the state of Baja California, the major features of the rift system are the Guaymas lineament, the north-trending zone of seismic activity, and the Cerro Prieto lineament (Fig. 55). The Guaymas lineament is believed to be a right-lateral transform-fault system (Phillips, 1966) connecting spreading centers in the southern half of the gulf with spreading beneath the northern half of the gulf. The Cerro Prieto lineament is a system of right-lateral transform faults that serves to connect the spreading centers of the northern gulf with spreading centers located beneath the Mexicali Valley of Baja California and the Imperial Valley of southern California.

Merriam (1965) applied the name "San Jacinto" to the faults that cut recent deposits of the Colorado River Delta and reach the Gulf of California just east of the mouth of the Colorado River at El Gulfo (state of Sonora). These faults are not in line with the San Jacinto fault of Riverside and San Diego Counties, California, and mapping has not connected the two trends. Barnard (1968b) referred to the fault that runs beneath Cerro Prieto as the "Cerro Prieto fault." We have combined these features in the "Cerro Prieto lineament" (Fig. 55). East of the Colorado River estuary, modern scarplets of the Cerro Prieto fault system are

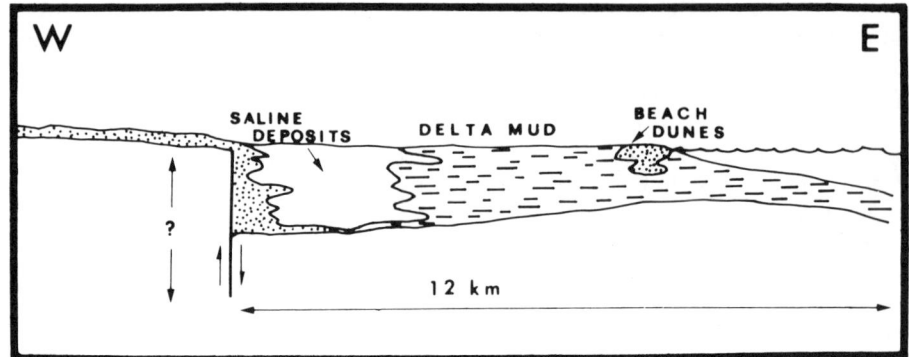

Figure 60. Hypothetical cross section across the gulf coastline north of San Felipe (loc. 13, Pl. 3).

up on the northeast side; this suggests that the lower, southeast-trending reach of the Colorado River lies in a subsiding graben. In the Cerro Prieto geothermal field 40 km southeast of Mexicali, however, the basement is down to the northeast (Alonso and Mooser, 1964).

The southwestern edge of this graben is presumed to be the fault that bounds the northeastern edge of the Sierra del Mayor. This fault trend can be followed to the southeast, where it forms the western boundary of a large mud flat bordering the Gulf of California (Fig. 11). The western edge of the mud flat runs straight northwest for 30 km and is exactly parallel to the western edge of the Laguna Salada and the eastern fault of the Valles San Felipe and San Pedro (Hamilton, 1971, his Fig. 9). We interpret the western shoreline of the mud flats as a depositional boundary determined by a fault that is downdropping the block on the eastern side. Figure 60 suggests that saline deposits occupy a persistently subsiding basin on the downthrown side of the fault.

Sierra de los Cucapas and Northern Sierra Pinta. The topography of the Sierra de los Cucapas (loc. 3) displays a pronounced northwest fabric that is shared by the distribution of metamorphic rocks, the foliation of the granitic rocks, the Cenozoic faults, and the Mesozoic intrusive contacts (Barnard, 1968b). This orientation parallels the trends of the San Andreas and Cerro Prieto faults.

McEldowney (App. 1) described 95 faults in the northern Sierra Pinta (loc. 12). Although the fault pattern is particularly complicated in this area, the most prominent faults trend N. 15° W. and N. 40° E., similar to those in some areas farther south. The N. 65° W. striking faults of the Sierra Pinta are only prominent north of Mexican Highway 5.

The Laguna Salada occupies the area between the Sierra de los Cucapas, Sierra del Mayor, and the main gulf escarpment to the west and the Sierra Pinta to the south. Kelm (App. 1) called this a graben on the basis of a gravity and magnetic survey. The topographically lowest part of the basin, the north-northeastern edge, overlies the deepest sedimentary fill as interpreted from gravity measurements. This suggests that the basin is still subsiding.

Southern Sierra Pinta, Sierra San Felipe, and Sierra Santa Rosa. The structural province that includes the southern Sierra Pinta, Sierra San Felipe, and Sierra Santa Rosa extends from the main gulf escarpment east to the gulf and from the northern Sierra Pinta on the north to Valle San Fermin on the south. It is characterized by intermountain basins and northeast-trending faults.

The most prominent basin in the province is the Valle San Felipe graben that separates the Sierra San Pedro Mártir from the Sierras San Felipe and Santa Rosa. This graben is in some ways analogous to that which separates the Sierra Nevada from the Inyo and White Mountains of California, U.S.A. Slyker (App. 1) studied the Valle San Felipe by gravity and magnetic surveys as well as by geologic mapping. He calculated the fill in the valley to be 2,400 m thick. If this indicates the depth to basement, there has been at least 5,300 m of uplift along the San Pedro Mártir fault.

The movement on the San Pedro Mártir fault must be continuing, because the fault trace is at the base of the bed rock with a relatively flat fan slope. Toward the southern end of Valle San Felipe (Valle Chico; loc. 17), recent fault scarplets cut the alluvial surface (Figs. 49 and 50). Just north of Rancho Parral (loc. 19) at the southern end of the Sierra San Pedro Mártir is an exposure of the dip of the faults that compose the main gulf escarpment (Fig. 61).

The eastern side of the Valle San Felipe graben is formed by the San Felipe fault (Slyker, App. 1). It has no surface expression in Valle San Felipe but is indicated by the gravity survey. Its southern extension continues down the center of the valley. Here, it is marked by sag ponds and truncated drainages (Fig. 57; 10 km north of loc. 17). Neither side appears to be rising, and no evidence is seen for strike offsets. This fault may result directly from the dilation of the valley.

The desert mountains that lie between the Valle San Felipe graben and the gulf coast basins have experienced a complicated Cenozoic history not shared by the mountains to the west. Volcanic rocks were deposited in the Eocene, Miocene, and Pliocene Epochs. Differential uplift and erosion have been taking place since at least early Miocene time, with sediment accumulating in the intervening fault basins. Marine incursions occurred in late Miocene and Pliocene time. The basins were folded between 9 and 6 m.y. ago (Sommer and Garcia, 1970). The monolithic megabreccia in the marine Pliocene strata northwest of San Felipe attests to the relief of the southern Sierra San Felipe at that time.

Puertecitos. The relatively small Puertecitos structural province, immediately adjacent to the gulf and extending onto the small nearshore islands, is mainly covered by Pliocene and Pleistocene volcanic rocks. Beneath the nearly flat lying

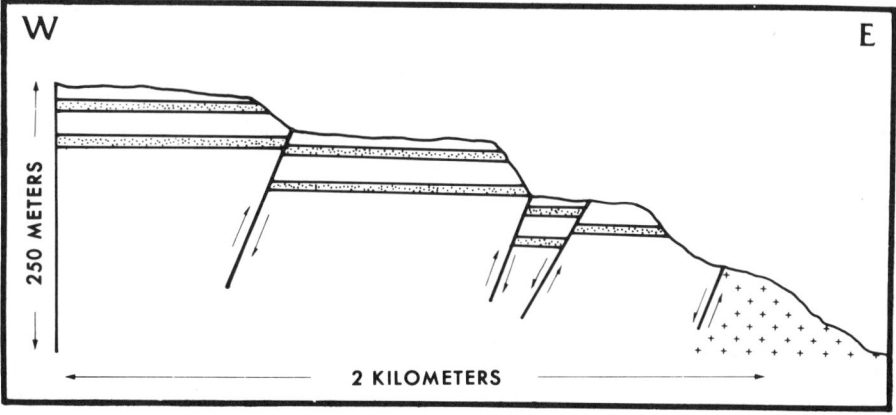

Figure 61. Diagrammatic sketch of a fault set north of Rancho Parral near the southern end of the Sierra San Pedro Mártir (loc. 19, Pl. 3).

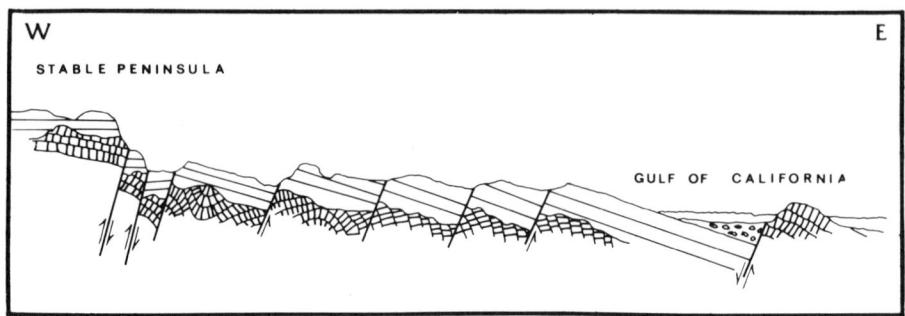

Figure 62. Generalized sketch of antithetic fault sets east of the stable peninsula in the Puertecitos structural province (province VIII in Fig. 55).

volcanic cap, the older Miocene volcanic rocks are tilted as much as 90° (Fig. 47). The north-trending fault pattern represents late Pliocene or Pleistocene faulting. Many of the faults are so uneroded that the full length of the fault scarps is clearly exposed. Nearly every fault in this region displays upward displacement to the east; this is antithetic to the depression of the gulf (Fig. 62).

Similar antithetic fault blocks between the main gulf escarpment and the gulf basin have been recognized in the Sierra Tinaja (loc. 11), the coastal areas between Bahía San Luis Gonzaga and Bahía de los Angeles (loc. 22), the Bahía de las Animas area (Loc. 27), and Santa Rosalía in the territory of Baja California (Wilson, 1948).

Rossetter (App. 1) investigated Isla San Luis (loc. 21; see also Fig. 36), the largest and most southerly of the six small volcanic islands between Huerfanito and Bahía San Luis Gonzaga. The majority of crevasses and other structural lineaments on the island trend north, parallel to the prominent onshore faults. It is possible to measure directly the sense of movement between the two sides of a crevasse in the most recent obsidian dome by matching vesicles that split when the sides pulled apart. The crevasse has dilated about 130 cm due east, normal to its strike. Points on the west wall are about 25 cm below the corresponding points on the east wall, consistent with the displacement of onshore faults. The islands and Cerro León, a large crater about 9 km southwest of Puertecitos, lie along the northwest extension of the Guaymas lineament (Fig. 55). Farther south in the Ballenas Channel, the lineament is characterized by a deep, narrow depression (the Salsipuedes basin). Phillips (1964b) believed that the lack of isostatic equilibrium and lack of appreciable sedimentary fill attest to the youth of the Salsipuedes basin. Where the lineament approaches the peninsula, the basin apparently has been filled by volcanic material. Dehydration dating of the youngest obsidian dome on Isla San Luis suggests that the rock may be less than 100 yr old. Thus, the basin and the volcanic ridge appear to be parts of an active system.

Gonzaga-Remedios Coastal Ranges and Basins. The Gonzaga-Remedios structural province lies between the stable peninsula and the Guaymas lineament. At its northern end, the province is clearly separated from the stable peninsula by the main gulf escarpment, but through the central area, the mountains of the unstable belt stand at the same elevations as the western and central portions of the stable peninsula. From Sierra la Asamblea (loc. 23) to Bahía San Luis Gonzaga, there is no clear-cut western boundary to the gulf depression.

Beginning at the southern end of the Puertecitos structural province and extending

30 km due south is a basin bounded by Mesozoic basement rocks on both the west and east sides. The basin contains Miocene andesite overlain by lacustrine and shallow marine strata of Pliocene and Pleistocene age and some younger volcanic strata. West of Bahía San Luis Gonzaga, the alluvial plain is crossed by scarplets that suggest recent fault activity.

For about 35 km north from Agua Amarga (loc. 25; see also Fig. 37), there is an irregular system of youthful faults forming three interior basins. We call these the Llanos de Amarga basins.

The gulf coast closely parallels the Salsipuedes basin for more than 50 km, and there are several faults that parallel the coast. No detailed mapping has been done in this province, but it obviously has undergone repeated volcanism, differential uplift, and erosion.

Isla Angel de la Guarda and Isla San Lorenzo. The Islas Angel de la Guarda and San Lorenzo and probably the ridge that connects them are underlain by granitic and metamorphic rocks. Most of the exposures on Isla Angel de la Guarda are tilted volcanic strata probably of Miocene age and capped by flat-lying volcanic strata probably of Pliocene age.

The northwestern part of Isla Angel de la Guarda shows a fault pattern with northwest and north-northwest trends; the southeastern part of the island has faults with north trends. Phillips (1966) found striking similarities between the structure of Isla Angel de la Guarda and the adjacent coastline; this suggests about 30 km, of right-lateral strike-slip separation.

Los Angeles–Las Animas. The complexly faulted area of Bahiá de los Angeles and the Animas basin is dominated by north-trending horsts and grabens. The Animas basin (loc. 27) displays a number of recent fault breaks and some interior drainage. The long volcanic history and varied structure of this small province place it high on the priority list for detailed mapping in the peninsula. The southern edge of the province is marked by an east-west boundary of unknown structure. Northwest-trending faults cross this boundary, and new faults parallel it. To the southeast, a relatively untilted granitic terrane extends to the gulf coast.

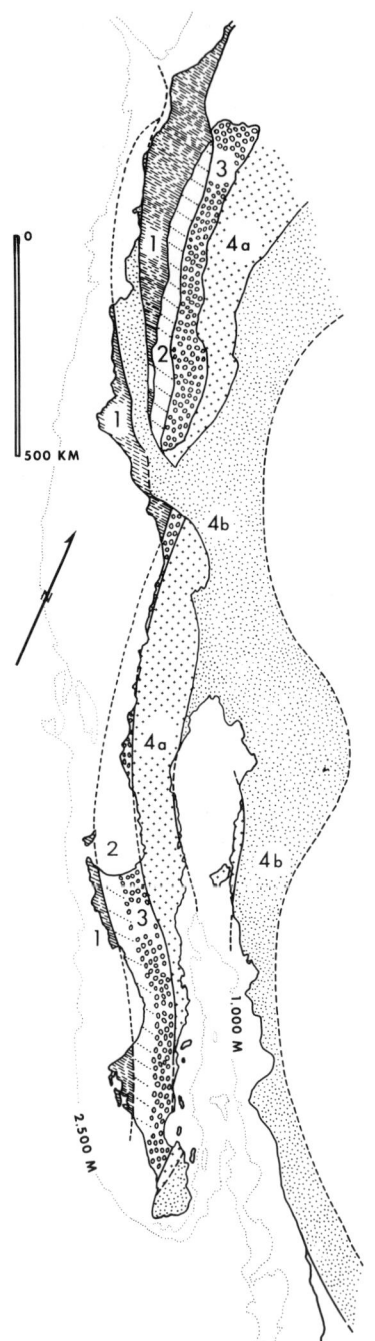

Figure 63. Structural framework of California, U.S.A., and Baja California. 1, Franciscan assemblage; 2, thick fill of the older Great Valley sequence; 3, thick fill of younger Great Valley sequence; 4a, Mesozoic-Paleozoic volcanic-volcaniclastic strata intruded by potassium-poor plutons; 4b, Paleozoic carbonate and quartzite strata intruded by relatively potassium-rich plutons.

11
Regional Tectonic Pattern

MESOZOIC ARC

We assume that the great cordilleran batholiths of North America were emplaced at the roots of dacite-andesite arcs analogous to those active today in Peru and Chile (Hamilton, 1969a). If we examine such an arc system, we find (1) an outer belt of ocean-floor and trench deposits (ophiolite, chert, graywacke, and metamorphic minerals such as glaucophane), (2) a belt of deep-water or continental-slope deposits (volcaniclastic rocks, graywacke, argillite), (3) a belt of synorogenic detritus of metamorphic and plutonic provenance, and (4) beneath the area once occupied by the arc of andesitic volcanos, plutonic rocks of gabbroic to granodioritic composition. Moore (1959) and Dickinson and Hatherton (1967) suggested that this belt of plutons becomes progressively potassium-rich toward the concave side of the arc.

In northern California, this first-order tectonic pattern (Fig. 63) is represented by the following belts: (1) Franciscan and associated rocks, (2) the older (Jurassic) part of the Great Valley sequence, (3) the younger Great Valley sequence, and (4) the Sierra Nevada batholith. The Sierra Nevada batholith can be further divided into (4a) a subbelt in which the country rock consists of Jurassic and Carboniferous volcaniclastic deposits that range from deep water to nonmarine and that are intruded by predominantly potassium-poor plutons of Jurassic and Early Cretaceous age and (4b) a subbelt in which the country rock is early to middle Paleozoic metasedimentary carbonate rock and quartzite that were intruded by relatively potassium-rich plutons of Late Cretaceous age. Similar lithologic patterns can be seen in southwestern Alaska and in other eroded arc systems.

South of the Transverse Ranges, the total picture is obscured because of the broad submerged area along the continental margin. Fragmentary evidence is present, however, in the Channel Islands, Islas Cedros and San Benito, the Vizcaíno Peninsula, and the Bahía Magdalena area. For the peninsula, then, we can discern (1) a western belt composed of pillow basalt, chert, graywacke, glaucophane schist, and serpentinite; (2) a belt of Jurassic and middle Cretaceous rocks possibly analogous to the older Great Valley sequence and so far recognized only on Isla Cedros (Kilmer, 1969) and on the Vizcaíno Peninsula (unpub. data of D. B. McIntyre,

E. C. Allison, and others); (3) the Rosario Group, consisting of metamorphic and granitic detritus derived during the uplift of the Cretaceous mountains; (4a) country rock of Triassic to middle Cretaceous, deep-water, shallow-water, and nonmarine (predominantly volcaniclastic) deposits that is intruded by a variety of Lower Cretaceous plutonic rocks; and (4b) country rock of chert, graywacke, quartzite, and carbonate units that has yielded only Carboniferous fossils and that is intruded by plutonic rocks yielding ages as young as Late Cretaceous.

If the batholiths of western North America once formed an arc, as suggested by Hamilton (1969b), were they part of a continuous structure connecting the Idaho, Sierra Nevada, and Peninsular Ranges batholiths with southern Mexico? Three prominent complications in this scheme are the Gulf of California, an area of partially oceanic crust; the Transverse Ranges, an east-trending belt of varied rocks including some of Precambrian age; and the Salinia block (Fig. 1), a 500-km-long splinter of carbonate and quartzitic rocks intruded by granitic rock whose age and composition are similar to those of the eastern Sierra Nevada (Ross, 1972). Wegener (1928), Carey (1958), Hamilton (1961), Rusnak and Fisher (1964), Wilson (1965), Normark and Curray (1968), and others have proposed fitting the peninsula of Baja California back against the coast of Mexico. Hill and Dibblee (1953), Curtis and others (1958), and many others have tried to realign the Salinia block between the Sierra Nevada and the Peninsular Ranges. Crowell (1962), Silver (1968), Ehlig (1971), and others have been concerned with the displacement of the rocks in the Transverse Ranges. In every case, the reconstruction has been accomplished by right-lateral displacement on the San Andreas fault. Movements of about 260 km are adequate to almost close the deep-water portion of the Gulf of California (Larson and others, 1968) and to remove the Precambrian rocks from their Transverse Ranges location (Crowell, 1962), but about 560 km is necessary to move the Salinia block (Hamilton, 1969b) to a position southeast of Tehachapi, California, U.S.A.

TECTONIC FRAMEWORK

The major structural elements of the peninsula of Baja California (Figs. 1, 55) are, from west to east, the continental borderland, the Santillán y Barrera line, the stable peninsula, the main gulf escarpment, and the basins and ranges of the gulf depression. The continental borderland has thinner than average crust (20 km thick) and high heat flow (see Chap. 9). This belt is separated from the stable peninsula by the Santillán y Barrera line (Fig. 55), a straight structural boundary dating from Cretaceous time. The stable peninsula is bounded on the east by the main gulf escarpment, a structure of late Cenozoic age. East of this is the Gulf of California depression, an area of basins and tilted ranges. Within the southern half of the gulf depression, there are deep unfilled basins underlain by crustal sections more typical of the ocean than of the continent (Phillips, 1964a). The Gulf of California is crossed by many fault systems that trend N. 40° to 50° W. The most southerly of these is the Tamayo fracture zone at the mouth of the gulf. There are four more between lat 23° and 27° N.; north of there, the Guaymas and Cerro Prieto rise-transform system has been described in Chapter 10.

Several authors have postulated transpeninsular faults. Normark and Curray (1968) drew the San Benito–San Hipolito lineament through the peninsula. This portion of the peninsula has been mapped (Mina, 1957) and is known to be covered by nearly flat strata of Miocene age; therefore, any major fault through this area must be Miocene or older. The Santa Rosalía fault drawn by Rusnak and others (1964) and adopted by Normark and Curray (1968) passes through an area of undisturbed Pleistocene deposits in its western portion and a largely unmapped area in its eastern portion. There are several discontinuous faults of similar trend in this part of the peninsula, but the Santa Rosalía fault (if it exists) must be of pre-Pleistocene age. Faults extending from the Guaymas lineament into the peninsula are also hypothetical and would be at least of pre-Pleistocene age. The Agua Blanca fault ends completely at San Matías Pass (Rogers, App. 2), and no fault of that trend extends to the gulf in this area. The San Miguel fault may form a chain with the Calabazos, Vallecitos, Las Palmas, Rose Canyon, and Newport-Inglewood faults (Moore, 1972), but this connection is speculative (Wiegand, 1970). The Elsinore fault of southern California is commonly extended across the peninsula to the Laguna Salada. No geologist has ever succeeded in mapping the Elsinore fault through the area east of Mason Valley, and it is doubtful if there actually is a continuous fault.

Normark and Curray (1968, p. 1599) proposed early transcurrent faults whose movements were first left lateral (A on Fig. 64). This recalls both our suggestion (Chap. 10) that the possibly left-lateral motion of the Santo Tomás fault preceded the right-lateral motion of the Agua Blanca fault and Fife's (1969) suggestion of left-lateral displacement for the Cretaceous Rosarito fault. Could this hypothesis also apply to the San Jacinto and Banning-Mission Creek faults of southern California? If so, this could help account for the differences in apparent distance of movement between the southern and northern segments of the San Andreas fault. If the southern fault segments moved first in a left-lateral sense and then in a right-lateral sense, whereas the northern fault segment only experienced right-lateral movement, the net right-lateral offset would be greater along the northern segment.

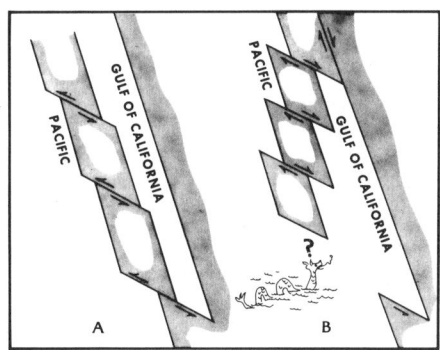

Figure 64. Interpretations of early opening of the Gulf of California: (A) after Normark and Curray (1968); (B) after Rusnak and Fisher (1964).

HYPOTHESES FOR THE ORIGIN OF THE GULF OF CALIFORNIA

Wegener (1928) was unaware of either the bathymetric or crustal nature of the Gulf of California. His only criteria were the shapes of the opposing coastlines. He clearly implied that the peninsula separated on the "San Andreas" fault, but beyond that, no mechanism was suggested. Carey (1958, his Fig. 24) conceived of the Gulf of California as a sphenochasm, much as Wegener had. Hamilton (1961) suggested both a strike slip of 560 km/100 m.y. and 160 km of rotation that swung the peninsula out from the mainland.

The real impetus for hypotheses of gulf opening came with the gravity survey of Harrison and Mathur; the bathymetric data of Rusnak, Fisher, and Shepard; and the seismic refraction studies of Phillips (all reported in *Marine Geology of the Gulf of California*, van Andel and Shor, eds., 1964). The work of Menard (1960) on the East Pacific Rise and the heat-flow studies of Von Herzen (1963) also added support. With these data, it became clear that there is oceanic crust under large areas of the southern half of the gulf and that these areas could not have resulted from simple subsidence of the continent. The deep crustal structure and composition of the northern gulf remain ambiguous. Deep seismic soundings have been unable to obtain the depth to the mantle, the velocity of seismic waves in the crust appears to be high and variable, and there is an apparent contradiction between gravity and seismic data.

Rusnak and Fisher (1964) presented a scheme by which the peninsula slid out like a deck of cards, the southern slices moving farther than the northern ones (Fig. 64B). By this hypothesis, the northwest lineaments of the gulf represent directions of crustal-block sliding, and the hypothetical transcurrent faults show right-lateral strike slip. The basins of the northern gulf were assumed to have thinned plastically. Phillips (1964b) presented a two-stage movement for the peninsula with a total displacement of 475 km. This restoration (Phillips, 1964b, his Fig. 7-6) aligns the continental escarpment west of Baja California with that of southern Mexico. The original "San Andreas" is envisioned as a continuous fault tracing the eastern edge of what is now the gulf, extending across the tip of what is now the peninsula (the La Paz fault), and then continuing west of what are now the Islas Tres Marías. Alternatively, the original fault passed inland east of the granitic rocks of the state of Jalisco (about lat 20° N.) and followed the trans-Mexican volcanic belt. The earliest movement, according to Phillips's hypothesis, removed the southern tip of the peninsula from the embayment that lies between the Islas Tres Marías and the mainland coast of Sonora. This portion of the gulf is underlain by a thin crust and a thick prism of what might be well-indurated sedimentary rocks or volcanic strata (seismic velocity of 5.1 km/sec). Except for the creation of this basin at the mouth of the gulf, the initial motion of the peninsula would have been simple right slip. This hypothesis achieves the best fit between the peninsula and mainland, the same fit that Wegener recognized.

In 1964, the motion of Baja California away from Sonora was conceived of as a motion of Baja California over Pacific mantle. With the advent of the theory of sea-floor spreading, however, the East Pacific Rise was classified as a spreading center, and the separation of Baja California from the mainland could be achieved by the creation of new ocean crust between the Pacific and North American plates without any relative motion between Baja California and the Pacific. Wilson (1965) conceived of the San Andreas fault as a rise-rise transform fault connecting the East Pacific Rise in the Gulf of California with the Gorda Ridge off the states of Oregon and Washington. Allen (1968), Larson and others (1968), and Moore and Buffington (1968) conceived of each of the en echelon northwest-trending lineaments of the gulf as a separate transform fault linked by a short segment of spreading center. In the southern gulf, these short rise segments correspond to northeast-trending topographic features located within the deep basins. Under the Colorado River Delta and the Imperial Valley, the short rise segments are assigned to the areas of abnormal thermal activity.

Larson and others (1968) used the record of the magnetic anomalies just within the mouth of the gulf to interpret the amount and rate of gulf opening. They discovered that anomalies dating back 4 to 6 m.y. B.P. are found between Cabo San Lucas and the Islas Tres Marías. They interpreted this to mean that the mouth of the gulf has dilated 260 km in 4 to 6 m.y.

There are many reasons to believe that the Gulf of California existed much earlier as a structural depression and a marine seaway (Karig and Jensky, 1972; Moore, 1973). K-Ar dates (Chap. 5) imply that gulf volcanism began at least 28 m.y. ago, and there is reliable structural evidence that the formation of the gulf was associated with this volcanism. Near San Felipe, diatomaceous marine Miocene strata have been found (Andersen, App. 1); in western Sonora (Gomez, 1971), the lower(?) Pliocene Imperial and Salada Formations are distributed widely throughout the gulf depression. Larson and others (1968) and Normark and Curray (1968) suggested that the gulf opened first in the north, having swung out along left-lateral transcurrent faults. This would have allowed cold-water California coastal species to enter the gulf, but gulf faunas have always had warm-water "Panamic" affinities (Stump, App. 1).

DISCUSSION OF RECENT HYPOTHESES

The necessary evidence for proof of fault displacement is the recognition of two separated rocks that were once united. Thus, if we are to reconstruct the Mesozoic arc as illustrated by Hamilton (1969b), we have to be sure that there is good reason to believe that the parts of the puzzle are indeed out of place. Salinia, the block of intensely metamorphosed carbonate and quartzite units intruded by Cretaceous granodiorite and now found between the Franciscan belts of western California, U.S.A., does not fit the tidy structural pattern of Mesozoic circum-Pacific orogeny outlined earlier in this chapter. Its anomalous position was the original reason for proposing large-scale (560 km or greater) movement on the San Andreas fault (Hill and Dibblee, 1953) and remains the principal evidence.

The existence in Salinia of metamorphosed Paleozoic or possible Precambrian rocks related to some pre-Mesozoic structural configuration would not in itself be unreasonable. Such rocks are found between the positions of the Mesozoic trench and the volcanic arc in southeastern Alaska and British Columbia and between the modern trench and the volcanic belt in southern Mexico, Peru, and Chile. However, the fact that the granitic rocks of western California, U.S.A., are found only within the Salinia block and the fact that the Salinia country rock, intrusive composition, and intrusive age all correspond to that of the eastern Sierra Nevada batholith constitute a remarkable coincidence. Nevertheless, it is only justifiable to slide the Salinia block in from elsewhere if the proposed original position makes *better* structural and historical sense. To slide Salinia into the continental borderland off southern and Baja California (Suppe, 1970) would only move it from its present structurally unreasonable position to another position equally unreasonable.

The use of the San Andreas as a transform fault between the Pacific and North American plates (Wilson, 1965; developed further by Atwater, 1971, and Silver, 1971) creates a new and independent rationale for displacement on the San Andreas fault as well as defining the time interval in which this displacement should have taken place. This arises from the fact that the pattern of magnetic anomalies on the floor of the Pacific Ocean suggests that, while the oceanic crust was spreading toward and descending under the west coast of North America, the entire Pacific plate was moving northeast. This required a right-lateral motion between the North American plate and the Pacific plate. Before the East Pacific Rise reached North America (30 m.y. ago), this right-lateral motion presumably followed the continental margin, but after the arrival of the rise at the Pacific coast, a portion of the North American plate became attached to the Pacific plate and the right-lateral motion took place within the continent along the San Andreas fault. With this assumption, the displacement of the past 30 m.y. can be calculated at about 1,800 km (Atwater, 1970, p. 3525). Atwater (1971) believed that the major part of this displacement could be attributed to the San Andreas fault, with the balance attributed to the deformation of the Basin and Range province and to other faults. This youthful age for displacement is in conflict with the earlier belief (Crowell, 1962) that displacement on the San Andreas fault system has been accumulating since the Cretaceous Period. The mechanics of plate-tectonic theory could also be satisfied by continued slip along the continental margin even after the San Andreas fault formed.

If we reconstruct North America according to the scheme of Hamilton (1969b), we find that the Salinian segment of the batholithic chain is very different from both that of the Sierra Nevada batholith to the north and that of the Peninsular Ranges batholith to the south. First of all, rock assemblage belt (4b) is in direct contact, across the Sur-Nacimiento fault (see the top third of the map in Fig. 63), with rock assemblage belt (1). Rock analogues of the western Sierra Nevada are entirely missing, and there is no structure analogous to the Great Valley (Fig.1) or Baja California synclines (see Chap. 10) with their great thickness of synorogenic and postorogenic sedimentary rocks. Page (1970) attempted to explain this discrepancy by suggesting that the Salinia block was thrust westward across belts (2), (3), and (4a). A second discrepancy arises in the fact that the Salinia block is overlain by Upper Cretaceous and Miocene marine strata, whereas deposits of these ages do not even approach belt (4b) in the Sierra Nevada or the Peninsular Ranges. The similarity of these Upper Cretaceous and Miocene strata to those found elsewhere north of the Transverse Ranges and the dissimilarity of these strata to any rocks found in northern Sonora, southeastern California (U.S.A.), or Arizona certainly suggest that these strata were deposited on the Salinia block after it arrived at a position near its present latitude, thus indicating Cretaceous and early Tertiary movement. Sliter (1968, p. 37-38, his Figs. 8 and 9) and Allison (1969) found the Late Cretaceous climatic zone of central California, U.S.A., to be distinct from that of southern and Baja California.

A final observation on current hypotheses deals with the relation of magnetic striping of the ocean floor to the opening of the Gulf of California (Larson and others, 1968; see Fig. 65). The idea that the mouth of the gulf spread during the formation of the young crust that underlies it is not a unique solution in light of evidence. The gulf could have been there when the East Pacific Rise arrived at Cabo San Lucas (the southernmost tip of the peninsula) 4 to 6 m.y. ago. Let us suppose that during the past 4 to 6 m.y. the rise has moved about one-third of the way across the gulf toward the Tres Marías basin (the northern terminus of the Middle America Trench) and that the 4- to 6-m.y. signatures that were created just east of Cabo San Lucas are now two-thirds of the distance (260 km) to the trench. By this hypothesis, the Tamayo fracture zone is continuous with the remarkable escarpment just west of the Islas Tres Marías, and the entire product of sea-floor spreading has been going down the trench as fast as it formed. The

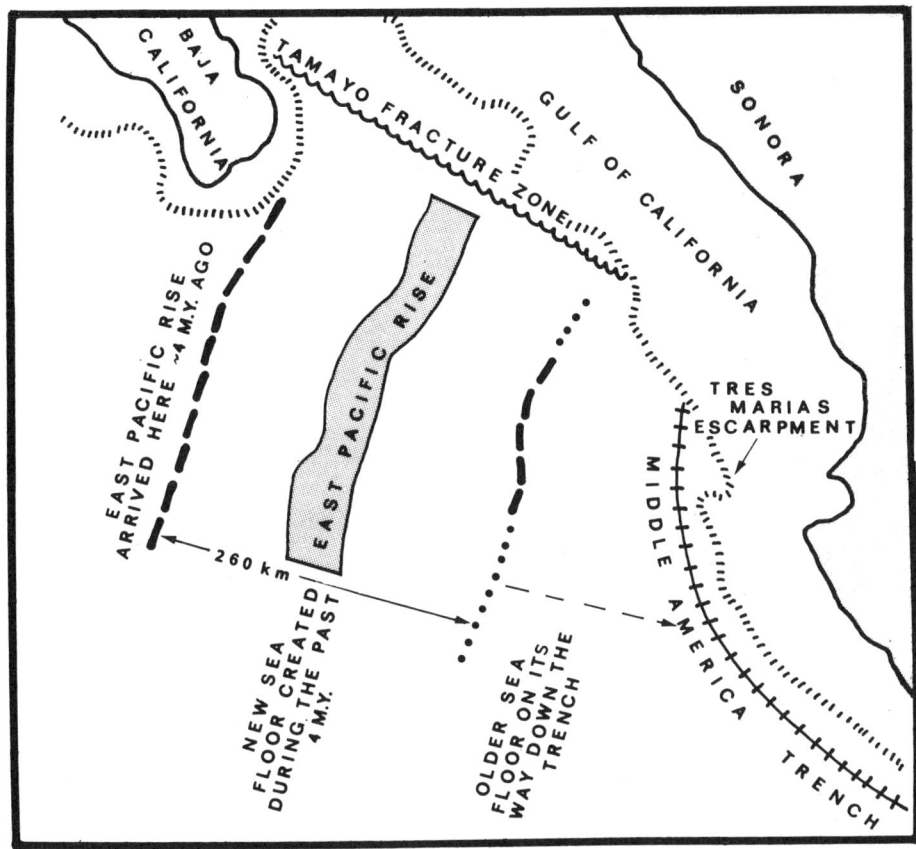

Figure 65. The East Pacific Rise at the mouth of the Gulf of California (after Larson and others, 1968).

identity of magnetic anomalies north of the Tamayo fracture zone is uncertain, making it impossible to assign a precise age to the new sea floor that is actually within the Gulf of California.

A successful reconstruction of Mesozoic western North America must reconcile the position at various times of four major tectonic elements: (1) the arrangement of the Paleozoic boundaries between quartzite and carbonate sedimentary rocks and wacke and chert volcaniclastic rocks, (2) the distribution of Jurassic volcanic rocks, (3) the distribution of middle Cretaceous volcanic rocks, and (4) the distribution of deep-seated plutonic rocks of Late Jurassic and Cretaceous age. To aid in reconstructing the events within the volcanic belts, it may be possible to recognize boundaries between marine and nonmarine facies; within the plutonic belt, it may be possible to recognize subbelts distinguished by differences in chemical composition and time of cooling. Our work in the peninsula and current mapping in Sonora and Sinaloa suggest that most of the above tectonic elements can be recognized on both sides of the Gulf of California (Gastil and others, 1972). We believe that the displacement of the Baja California Peninsula began in late Mesozoic time and that the gulf depression was already a seaway in Miocene time (Gastil and others, 1972).

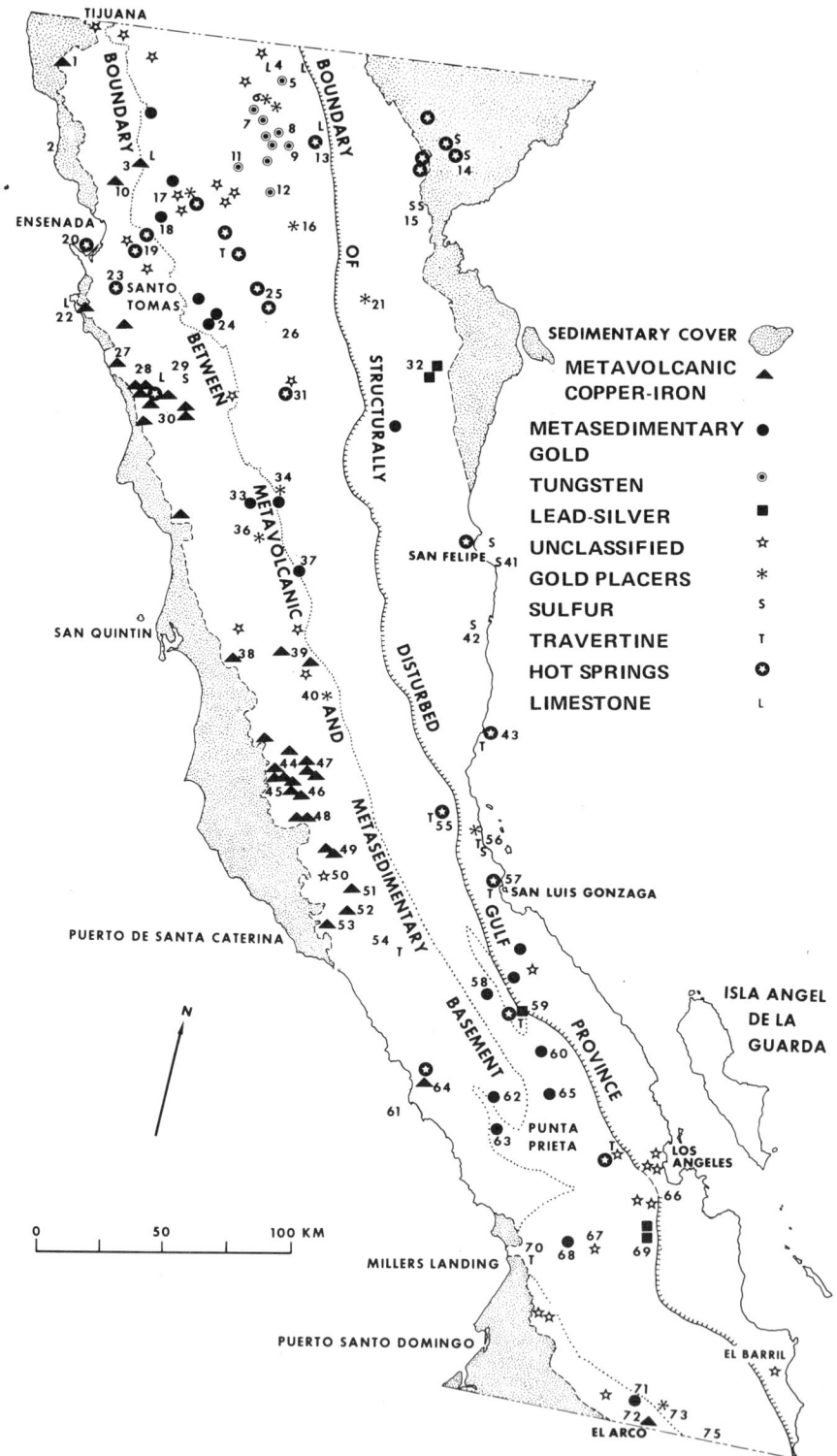

Figure 66. Mineral deposits in the state of Baja California. Numbers 1 through 75 correspond to localities cited in the text.

12
Economic Geology

HISTORY OF MINERAL EXPLORATION AND EXPLOITATION

During the mission period, from first exploration until the 1840s, official documents and diaries indicate that mining, except for salt and sulfur, was discouraged. The silver and gold mines in the mountains south of La Paz were the only reported mineral discoveries (Meigs, 1935; Aschmann, 1959).

There are, however, several reasons to believe that some exploration and mining did take place during this interval. For example, Aschmann related the following in a footnote (1959, p. 25):

When, in 1774, an Indian found a good-sized ingot of minted gold on the beach at Bahía de los Angeles, the missionary promptly reported it to the royal officials. It was decided that the bar had been lost by someone engaged in illegal operations and was, therefore, forfeit to the crown. The Dominicans accepted this judgment without argument. They appear to have been glad to lose the gold rather than arouse suspicion that they possessed a secret mine.

It is difficult to visualize how a mining operation so large that it could accidentally lose a good-sized ingot could have gone unnoticed in Baja California.

In 1792 José Longinos Martínez made the first expedition to the peninsula for the express purpose of recording the plants, animals, minerals, and other natural phenomena (Longinos Martínez, 1792). Although Longinos Martínez was a botanist by training, he was a keen observer of minerals and rocks.

He wrote the following account of Arroyo Calamajué (loc. 59; all locations mentioned in Chap. 12 are shown in Fig. 66):

In this canyon there are veins of silver and gold which have never been worked, doubtless because of the desert nature of the place and the want of every facility. Throughout the district there are hills which can be seen from a distance to be green with an abundance of *copperas* [p. 29].

The deposits that have since been mined are not apparent to cursory observation by simply riding through the canyon. The green that he saw was chlorite schist. He described the deposits on the western side of the peninsula (p. 13):

Near San Fernando [loc. 48] there are various veins of silver, lead, and magnetic iron (*piedra imán*). In the canyon of Santa Ursula there is a vein of hard iron (*hierro acerado*).

In the vicinity of this mission [Rosario] toward the high Sierra de Gentiles, one finds veins of iron and copper [loc. 47]. At the place called Las Cuevas, there is a hill with green lead (*plomo verde*).

At the foot of the high range which overlooks the valley of San Rafael [loc. 17], there are veins of iron, a kind of hematite. On its foothills and summits one finds only talc crystallized in different ways, quartz pebbles, and very rare rock crystals in large pieces. I made a reconnaissance of a great vein of iron near Mission San Vicente [loc. 28], some eight leagues in length. Between this canyon and that of Santo Tomás, on a high hill near the coast [loc. 27], there is a great deal of talc in sheets, mixed with hard iron (*hierro acerado*).

In various parts of the arroyo of San Vicente [loc. 28], there are veins of native alum and earths of different colors for paints, and, in the arroyo of hot water of this mission, veins of *copperas*.

All of these deposits have since been explored or exploited. In fact, Longinos Martínez managed to see a large proportion of the better mineralized areas that are known to exist in the peninsula. It is apparent that those who guided him from mission to mission were not unaware or unconcerned with mineral resources.

A case could even be made that most of the missions of central and northern Baja California were located near mineralized areas: Examples are San Borja (loc. 67; Au and Ag), Calamajué (loc. 59; Au and Ag), San Fernando (loc. 48; Ag and possibly Pb; Cu and Fe), San Vicente (loc. 28; Ag, Cu, and Fe), Santo Tomás (Ag and Cu), San Miguel (loc. 2; Ag and Cu), and Santa Catarina (Au and possibly Ag). This apparent correlation between mission sites and mineralized areas is probably only coincidental. The mission sites have had the longest history of occupation of any area in Baja California, and discovery of minerals has been related to habitation. The mission sites are also closely correlated with water supplies; many sites are near hot-water springs. It is possible that this persistent thermal activity has been the source of mineralization.

The apparent contradictions between the official ignorance and the obvious knowledge of mineral deposits during the mission era undoubtedly contributed to the legends of lost mines and lost missions and to the recurrent theory that some Baja California missions were engaged in mining precious metals. The weight of scholarly research into the history of these missions, however, finds no basis for such ideas. Soldiers and other adventurers probably found, and in some cases mined, the obvious showings of minerals on the peninsula during the mission era in spite of the missionaries.

When Gabb journeyed up the peninsula, the last mission had been abandoned in 1849, little more than a generation before. Gabb wrote (in Browne, 1868, p. 635):

Prospecting has been carried on over the whole length and breadth of the country, but, on the whole, without very marked success. . . . of gold mines there have been many, but at the present time not one is being worked. . . .

In the granite mountains from Sta. Gertrudis [loc. 75] to San Borja [loc. 67], and even in the metamorphic sandstones, almost as far south as San Ignacio, there are innumerable tunnels, shafts, and coyote holes, where attempts have been made to find paying quartz mines. They are now without exception abandoned. . . .

Similar attempts have been made to discover or develop silver mines. These are reported as existing about San Borja, on the island of the Guardian Angel, on the mainland opposite this island, on the Island of Margarita, and in numberless other places, none of which have ever yielded anything nor probably ever will. . . .

Copper, like gold, is reported from nearly every part of the territory; numberless mines have been opened and invariably abandoned.

The only metal mines that he found in operation were the silver mines of the Triunfo district south of La Paz and the Delphina copper mine near San Vicente (loc. 28).

Ironically, Gabb's dreary appraisal was ill-timed; unknown to him, in the previous year one of the world's most remarkable copper deposits had been discovered at Santa Rosalía, and gold was just being discovered at Calmallí (loc. 71). Within another generation, gold and silver discoveries, some in the areas to which he alluded, were to transform the economy of the peninsula. The ports of Ensenada, Bahía de los Angeles, Miller's Landing, Santa Caterina Landing, Punta Baja, Bahía San Luis Gonzaga, Puerto Santo Domingo, and the roads from these points to the interior were all constructed to bring supplies to the mines. A few of the mining towns that flourished during this era are Real del Castillo (loc. 17; for a time the capital of the territory), El Alamo (loc. 24), San Fernando, Las Flores (loc. 66), Punta Prieta, Calmallí, Pozo Alemán (loc. 73), and El Arco (loc. 72; see Zárate and Nuñez, 1925). Discoveries continued into the early part of this century (Zárate, 1922), but most properties were abandoned by 1929. There was some resumption during the 1930s, but by 1949, only one small gold mine at San Ignacitos (loc. 68) was in operation (Aschmann, 1959). An interesting report on the mineral deposits in the vicinity of Calmallí (loc. 71) was made by Ramos (1885).

MINERAL PROVINCES

The peninsula is divided into five mineral provinces (Fig. 66). The westernmost province contains mesothermal iron and copper sulfides and iron oxides, emplaced in metavolcanic rocks. Gold and silver values are commonly present. The province extends from Orange County to El Arco at lat 28° N. Deposits are disseminated or in veins and are more or less related to granitic intrusive rocks, particularly those of adamellite or granodiorite composition. The restriction of these deposits to areas of metavolcanic basement rock and the prevalence and wide distribution of copper suggest that the ore has been leached from the country rock by sulfur-bearing solutions mobilized by metamorphism.

The second province contains gold lode deposits in metasedimentary rocks. This province follows the axis of the peninsula from the Julian district in San Diego County through the Sierra Juárez, Sierra San Pedro Mártir, Sierra la Asamblea (or San Luis), and Sierra San Borja to the Calmallí district. These deposits commonly occur near the boundaries of plutons, particularly those emplaced at intermediate depths. The common association of these deposits with metamorphosed detrital sedimentary rocks suggests the possibility that they have been reworked from ancient placers.

In the third province, deep-seated contact-metamorphic deposits in carbonate-bearing metasedimentary rock are found in the northern portion of the stable peninsula. The principal commodities of this province have been tungsten and precious stones.

The fourth province is of Cenozoic age and corresponds to the area of Cenozoic volcanism found primarily in the gulf depression. It includes many surface deposits of travertine, near-surface deposits of sulfur and manganese, and slightly deeper deposits of copper, silver, and lead sulfides, wulfenite, stibnite, and other minerals. This type of mineralization is continuing today, as shown by the highly mineralized brines of the Niland geothermal field of the Imperial Valley, California (Muffler and White, 1969) and by the numerous sulfurous and carbonate hot springs still active in the peninsula and beneath the Gulf of California. It is at present impossible to ascertain whether some of the sulfide deposits in the basement rocks of the eastern peninsula are of Mesozoic or Cenozoic age. For the purpose of discussion, we assume that deposits primarily of gold are Mesozoic, whereas those primarily of lead and silver are Cenozoic.

A fifth province contains the gold placer deposits that initially accumulated on the old erosion surface of early Cenozoic age.

Volcanic Copper-Iron Province

The most complete annotation of these deposits is by Böse and Wittich (1913). Just west of Valle Cuero de Venada is the mine El Sueño (loc. 1) where an openpit mine and a modern mill were in operation in 1963. They closed soon thereafter. Wittich visited El Sueño, Santa Rosa (somewhere between La Misión and Rancho Santa Rosa), and San Antonio (loc. 10) in 1911. He found only fracture coatings of copper carbonate. He reported that copper (chalcopyrite) was shipped from Mina de Marguerite (loc. 22) located near Punta China south of Valle Santo Tomás. He found additional prospects near San Antonio del Mar (loc. 30). Some of these contained chalcopyrite, cuprite, and native copper. Mina San Mateo near Punta San Isidro (loc. 27) west of San Vicente exposed only small amounts of copper carbonate. Mina Delfina (loc. 28) near Misión San Vicente was visited by Gabb in 1866 (in Browne, 1868, p. 636). The mine was apparently shipping ore from the oxide zone at that time. Wittich found the remains of a small copper smelter when he visited Mina Delfina in 1911.

Four kilometers from Misión San Vicente is a deposit of hematite iron ore called Cerro Colorado that Wittich (1915) estimated at 5 million tons. According to Wittich, it is a hot-springs deposit, but others who have seen it believe it to be related to contact metamorphism. Wittich's analyses are reported in Appendix 10. Mina Tepuztete, an iron mine near Punta San Isidro (loc. 27), is located at the intrusive contact between granitic and carbonate rocks.

In 1964, we found evidence of mining in the basement rocks east of Valle Santa María (loc. 38) and in the vicinity of Rancho El Rosarito (loc. 39) near the southern end of the Sierra San Pedro Mártir. Small-scale smelting had been done at Rancho El Rosarito.

Between Rancho Aguajito and El Arenoso, east of El Rosario, there are a number of localities in which exploration for copper and iron has been carried out (Consejo de Recursos Naturales no Renovables, 1965). Mina Turquesa (loc. 44) has been mined intermittently for turquoise; Mina Caramayola (loc. 46) displays only chrysocolla. Mina Sauzalito (loc. 47) shows a mixture of iron oxides and copper carbonates.

Over this entire area, copper-stained rocks are not uncommon. South of El Arenoso, Mina Ursula (loc. 45) appears to contain good-quality, hard hematite ore. A hilltop estimate suggests that there might be as much as one million minable tons.

South from Rancho Progresso are the adobe ruins and waste dumps of the San Fernando district (loc. 48) and the nearby El Gato, also called "Mile of Iron." Wisser (1954, p. 68) reported that the San Fernando ore consists of chalcopyrite disseminated through "diabase."

Along a trend of S. 45° E. from San Fernando are the Evolución and Esmeralda mines (loc. 49), Mina Luciano (loc. 51), and Mina Evangelina (loc. 52). Farther south and on a different trend are Santa Catarina, Julio César, and Cleopatra mines (loc. 53). South of Puerto de Santa Caterina, there is little evidence of the province. A claim for hematite, however, is located in Mesozoic volcanic rocks just north of Punta Cono (loc. 61). El Arco (loc. 72) is apparently the southernmost district that belongs to this province. The American Smelting and Refining Company of Mexico has recently outlined a large low-grade copper deposit at El Arco.

Significantly, Longinos Martínez (1792), Gabb (in Browne, 1868), Wittich (1911, 1915), and Wisser (1954) reported the same deposits. The small iron deposits probably would have been exploited except that Mexico has reserved iron ore for use within Mexico; the industry of Baja California is not yet developed enough to use iron ore. The copper deposits are too low grade for production on a small scale in such remote areas. If large low-grade deposits exist, the province may be ready for a new and more profitable era of mining.

Schist-Gold Province

These deposits have been mentioned by Lindgren (1889), Böse and Wittich (1913), Flores and González (1913), and Wisser (1954). The most northerly district of significance is just north of Real del Castillo (loc. 17). When this area was visited in 1963, a small amount of development and maintenance work was continuing, but by 1971 it had been abandoned. The rocks closely resemble the Julian district of San Diego County, California (Weber, 1963) as do those at Mina las Cruces (loc. 18) between Valle San Rafael and Ensenada. At El Alamo (loc. 24) and at the Peterson mine west of the Sierra San Pedro Mártir, the gold is in quartz veins that cut tonalite bodies lying near the boundary of the metavolcanic and metasedimentary basement terranes. At Socorro (loc. 34), there are auriferous quartz veins cutting metasedimentary gneiss. Between Bahía San Luis Gonzaga and Arroyo Calamajué (loc. 59), many gold prospects have been explored, and considerable money has been spent on roads, mills, and development work. Gold has also been mined in the metasedimentary rocks near Laguna Chapala (loc. 58), at the contact between slate and tonalite on the western slope of Cerro San Luis (loc. 60), at Desengaño (loc. 65), at León Grande (loc. 62), at Mina Columbia (loc. 63) west of Punta Prieta, and in the hills north and west of Bahía de los Angeles. In several areas, including Calmallí, copper sulfides occur with the gold.

Scheelite Deposits of the Carbonate Province

The known deposits of this province are restricted to the northern Sierra Juárez and have been described by Fries and Schmitter (1945) and by A. H. James (App. 2). Major mines are located at La Olivia (loc. 9), La Pelita, Los Gavilanes, and

El Fenómeno (loc. 12). The deposits are generally related to the intrusive contacts between granitic rock and discontinuous lens-shaped bodies of carbonate rocks. Similar bodies found between the Sierra de los Cucapas and Sierra del Mayor in the southwestern Sierra Pinta and the eastern Sierra San Pedro Mártir have not yielded tungsten deposits. Böse and Wittich (1913, p. 383) mentioned the existence of tungsten in the Sierra San Pedro Mártir but gave no localities. Favorable rocks would seem to exist at many places in the central and eastern peninsula south of the Sierra Juárez.

Cenozoic Hydrothermal Province

Longinos Martínez (1792, p. 5) reported the following in his journal:

. . . about ten leagues to the east of Mission San Ignacio [territory of Baja California] there are several springs of water at a temperature of 58°, heavily charged with sulphur and vitriolic salt, which are not used for anything.

In the vicinity of the abandoned mission of Calamajué [loc. 59], there is a spring on the slope of some fairly high hills, on some layers of earth (*tepetates*) or platforms composed of the abundant selenite of this water which has become incrusted around its circumference. From the center of this incrustation flows a stream the thickness of a tile and almost cold (10°), acidulated with vitriolic acid and aluminous selenite (*selenita aluminosa*). The passers-by, principally soldiers, have the custom of bringing sugar and drinking this water as if it were lemonade. They say it is refreshing, but they are mistaken, for its effects are quite the contrary.

Near the pueblo of San Luis [loc. 57] at the foot of the hills of Santa María (abandoned mission), there is a spring which, after it has flowed a little way, deposits a purgative neutral salt (*sal neutra*), similar to Glauber's salts. In its vicinity other springs of the same sort deposit marine salt in foamy incrustations.

About six leagues to the east of Mission San Vicente, among the Cerros de Fertilidad, there are several water holes in an arroyo which are called Agua Calicite [loc. 29]. All of them together yield a flow of some eight to ten oranges at 70° Réaumur. The water flow is charged with ferric sulphate (*vitriolo de Marte*). It is not used because it is out of the way and among gentiles.

Near the port of San Felipe de Jesús, on the beach next to the water, there is a hot spring of some size which can only be seen at low tide. If any fish passes near the spring at high tide it is killed or stupefied.

At the place called La Grulla, between Mission Santo Tomás and Mission San Miguel there are several springs in a fairly wide plain.

There are literally hundreds if not thousands of these thermal and mineralized springs. Examples not mentioned by Longinos Martínez are Agua Caliente near Tijuana, Marconi east of Real del Castillo (loc. 17), Ramírez and San Carlos (loc. 19) east of Maneadero, Punta Banda (loc. 20), Cañon Virgen de Guadalupe in the Sierra Juárez (loc. 13), San Vicente (loc. 28), Arroyo Volcán below El Mármol (loc. 55), the travertine deposits north of Bahía San Luis Gonzaga (loc. 56), Cerrito Blanco (loc. 54), Las Palomas (loc. 64) near the southwest coast, and the thermal area of Cerro Prieto (loc. 14) south of Mexicali. Böse and Wittich (1913) indicated thermal springs in Arroyo Guadalupe north of Real del Castillo, Escalullas north of El Alamo (loc. 24), and Valle Trinidad (loc. 31). A few of the springs, for example, those along the Agua Blanca (locs. 20 and 31) and San Miguel (loc. 25) fault zones, could be related to fault action, but others appear unrelated to either faults or recent volcanism.

We believe that there is a definite relation among the hot springs, the travertine deposits, the sulfur deposits, the metalliferous brines of the Niland geothermal field of the Imperial Valley in southern California, the Boleo copper deposit of Santa Rosalía and large manganese deposits of Santa Rosalía and Bahía Concepción, the manganese deposits found near Blyth along the Colorado River, the lead-silver occurrences of Tertiary age, and possibly even the remarkable pure gypsum deposits of the Imperial Valley and Bahía Concepción.

Most of the modern springs are only known to be depositing $CaCO_3$ and emitting CO_2. Native sulfur is deposited in bedrock fractures below the travertine deposit at Miramar (loc. 56). Barite has been deposited around the hot springs of Puertecitos (loc. 43), and sulfurous fumes are reported from the beachline springs at Punta Diggs (loc. 41) and offshore springs at San Felipe. Recent bulldozer excavations at Agua de Chale (loc. 42), a sulfur mining area, show the intricate relation of sulfur deposition to surficial siliceous sinter deposits. Beneath these sinter deposits, sulfur and coarsely crystalline quartz have replaced the granitic country rock. Sulfur is also deposited by hot springs in the Cerro Prieto (loc. 14) area and between the Sierra de los Cucapas and Sierra del Mayor (loc. 15).

Many of the active travertine deposits show the deposition of small amounts of MnO_2 and Fe_2O_3 along with $CaCO_3$. Flores and González (1913) reported an analysis (App. 11) from the hot spring on Arroyo Volcán, below the onyx mine at El Mármol (loc. 55); the mine was known earlier as New Pedrara. Alonso and Mooser (1964) reported an analysis (App. 12) for brine from well M-3 in the Cerro Prieto geothermal field. Muffler and White (1969) reported an analysis (App. 13) for brine from wells and mud pots in the Niland geothermal field, southern California. White and others (1963) pointed out that these thermal brines suggest that hydrothermal deposits of precious and base metals are now forming beneath the thermal areas of the Colorado River Delta. Muffler and White (1969, p. 169) reported the following:

Pyrite is detected by binocular examination of the cuttings at depths greater than 2100 feet and becomes conspicuous on the diffractograms at all depths greater than 3000 feet. Chalcopyrite and sphalerite are present sporadically from 2000 to 5232 feet, the total depth of the well. Chalcopyrite appears in the cuttings as discrete crystals and as veinlets cutting siltstone. Sphalerite occurs as irregular masses in sandstone but has not been seen in veinlets.

Doe and Tilling (1966) reported that lead and strontium in the brines were derived from the underlying Cenozoic sedimentary rocks.

The most important ore deposit in the peninsula of Baja California is at Santa Rosalía, lat 27°20' N. on the Gulf of California. Here, copper and accessory amounts of other elements occur in argillaceous marine tuffs of Pliocene age. Between 1885 and 1965, at least 15 different geologists have investigated the origin of this deposit (summarized by Wilson and Rocha Moreno, 1955; Nishihara, 1957). Hydrothermal, supergene, and a variety of syngenetic origins have been proposed. Each has been effectively criticized by subsequent authors, and no explanation yet presented is entirely satisfactory.

Wilson and Rocha Moreno (1955, p. 82) gave the following analysis:

> The elements that have a proportion many times higher in Boleo [Santa Rosalía] copper ores than their estimated averages in igneous rocks are copper, about 700 times as high; silver, 90 times as high; manganese, 65 times as high; zinc, 60 times as high; cobalt, 50 times as high; lead, sulfur, and chlorine, 40 times as high. All these elements are thought to have been introduced into the Boleo ores by solutions, although some of the chlorine may have been a residual product of sea water.

These data are presented in Appendix 14 (from Wilson and Rocha Moreno, 1955, p. 82, Table 25). The abundance of Mn, Zn, and Cl suggests a relation to brines from the Salton Sea area (App. 13). Adjacent to the Boleo deposit is the Lúcifer manganese deposit, which is similar to the strata-bound occurrence of the Boleo ore except that manganese rather than copper is the more important element.

The lead-silver Montezuma mine (loc. 32) at the southern end of the Sierra Pinta occupies a fault in Miocene volcanic rocks. Mapping in the northern Sierra Pinta revealed small amounts of copper and silver associated with chalcedony and opal in fractures and concretions. It is not known if the lead-silver deposits of Arroyo Calamajué (loc. 59) and Mina San Juan (loc. 69) south of Bahía de los Angeles are of Tertiary age, because the deposits are not in contact with post-Mesozoic rocks.

The marine Pliocene strata of the gulf contain several remarkably thick and pure deposits of gypsum, notably near Split Mountain west of the Imperial Valley in California, U.S.A., and near Bahía Concepción in the territory of Baja California. Touwaide (1930, p. 119) wrote:

> The origin of this gypsum has been attributed to precipitation by evaporation in a partly enclosed body of water or to precipitation from hydrothermal submarine springs. The writer inclines toward the latter theory, by mingling of sulphate water with sea water, since an abnormal local concentration of calcium sulphate by hydrothermal springs accounts best for the erratic and varied distribution of the gypsum bodies.

It may be significant that manganese oxide and strontium carbonate deposits are associated with the gypsum in the Fish Creek Mountains, western Imperial County, California.

Before steam wells were drilled, no one suspected from the CO_2 springs and mud pots of the Salton Sea that a very large reservoir of hot, metal-rich brines was below. Is it possible that the travertine deposits of the central peninsula overlie deposits of native sulfur, or that the sulfur deposits of Agua Chale or the Sierra de los Cucapas overlie sulfide deposits? It is conceivable that there is a depth zonation of such deposits: gypsum and travertine at the surface; sulfur and manganese dioxide just below the surface; wulfenite, argentiferous galena, bornite, and copper minerals at greater depths. If this is true, every hot spring, both active and ancient, suggests the possibility of a mineral deposit at depth. Looking at the Boleo copper deposit from the overview of the entire gulf structural province, it is surprising that no copper mineralization is found in the many exposures of marine Pliocene strata located elsewhere in the peninsula. It may be significant that Santa Rosalía is a locality in which once deeply buried Tertiary marine strata are adjacent to centers of post-Pliocene volcanism.

Placer Gold Province

The principal deposits of placer gold are at Campo Juárez (loc. 6), Los Piños, and Campo Nacional (loc. 16) in the Sierra Juárez; at Socorro (loc. 34), Valledares (loc. 36), and Los Enjambres (loc. 40) in the Sierra San Pedro Mártir; at Real del Castillo (loc. 17) on the north edge of Valle San Rafael; and at Pozo Alemán (loc. 73) near the southern boundary of the state. The discovery of placer gold generally preceded the discovery of lode deposits.

The deposits of Pozo Alemán, Real del Castillo, Socorro, and Valladares are all close to or overlie rocks of the schist-gold province; in each area, gold has been discovered in the adjacent bed rock. These placers probably are primarily the result of vertical accumulation by weathering and erosion; thus, the gold has not been transported far from its source.

The placers of the high Sierra Juárez, however, overlie granitic rocks and metasedimentary rocks that are not gold producers. Lindgren (1888) said that the gold at Campo Nacional was coarse and worn by stream abrasion. Minch (1970) determined by clast imbrication that all of the high gravel deposits of the Sierra Juárez were derived from the east. Some of the clasts are attributed to bed rock found along the gulf coast of the peninsula, whereas some others apparently came from Sonora.

OTHER RESOURCES

Petroleum. Much of the important geologic exploration of Baja California has been accomplished in the search for petroleum (work by Darton, G. D. Hanna, Beal, Santillán and Barrera, and Mina).

The summaries of Beal (1948) and Mina (1957) make a comprehensive treatment of this topic unnecessary. Despite a parallel geologic history and many stratigraphic similarities between southern California and Baja California, no indications of commercial petroleum have been found in the peninsula. Many exploration wells have been drilled in the southern half of the peninsula. The problem at most localities in the northern half of the peninsula is the thin, flat-lying nature of the marine sequence. West of the Santillán y Barrera line, however, there may be a much thicker stratigraphic section than those exposed on shore (geologic cross sections in Pl. 6).

In the Gulf of California depression are fault-bounded basins of late Miocene to Pliocene marine strata covered by thin layers of Pleistocene to Holocene alluvium (Pl. 6). These may provide reservoirs for petroleum.

Geothermal Energy. The geothermal field south of Mexicali, first noted by Bonillas and Urbina (1913) and since called one of the largest in the world (Facca, 1966), is now being developed. The field was described by Alonso and Mooser (1964), de Anda and Paredes (1964), de Anda and others (1964), and Mercado (1969). The distribution of thermal springs throughout the peninsula (Fig. 66), many of them near sea level, suggests that additional sources of thermal energy may exist.

Cement Limestone. At present, the most important mineral product in the state is limestone for the production of portland cement. This is mined from the limestone of the Alisitos Formation that crops out in a belt extending southeast from Punta China (loc. 22). The limestone is low in magnesium and aluminum. The large reserves and their proximity to deep-water shipping facilities make the area one of the most important sources of limestone for cement on the Pacific coast.

Marble outcrops in the Sierra Juárez and desert ranges are characteristically high in magnesium, but those near La Rumorosa (loc. 4) are mined for the manufacture of plaster. Small kilns have been operated at a number of localities west of the Laguna Salada and northeast of Valle Guadalupe.

Other Products. Other mineral products that have been explored or exploited include decorative and construction stone, perlite, pumice, mica, a variety of pegmatite minerals, precious stones, quartz, feldspar, corundum, wollastonite, barite, manganese, gypsum, diatomaceous earth, garnet, salt, and talc. Salt deposits were considered by Böse and Wittich (1913, p. 383-412) and Phleger (1969).

Appendix 1. Master's Theses

The following is a list of master's theses completed by students at San Diego State University concerning the state of Baja California. If any are not available through interlibrary loan, photocopies may be obtained through the Interlibrary Loan Office of San Diego State University.

Acosta, Marcial G., 1966, Geology of the Bahía Soledad embayment, Baja California, Mexico: 93 p.
Andersen, Robert L., 1973, Geology of the Playa San Felipe quadrangle, Baja California, Mexico: 214 p.
Bailey, Stephen M., 1966, Paleocurrent analysis of the Cretaceous Rosario Formation, Baja California, Mexico: 69 p.
Butler, Louis W., 1964, Geology of a submarine valley on the continental slope off Baja California, Mexico: 103 p.
Earl, John L., 1965, X-ray fluorescence rubidium: Strontium age determinations of minerals from the Southern California batholith: 122 p.
Fife, Donald L., 1968, Geology of the Bahia Santa Rosalia quadrangle, Baja California, Mexico: 100 p.
Flynn, Clinton J., 1968, Geology of the La Gloria-Presa Rodriguez area, Baja California, Mexico: 70 p.
Itson, Sonja P., 1970, The petrology of the mafic complex north of Jamul, San Diego County, California: 102 p.
James, Alton H., 1973, Structure and stratigraphy of the southern Sierra de Pintas, Baja California, Mexico: 56 p.
Kelm, Donald L., 1972, A gravity and magnetic study of the Laguna Salada area, Baja California, Mexico: 103 p.
McEldowney, Roland C., 1970, Geology of the northern Sierra Pinta, Baja California, Mexico: 78 p.
McGee, David C., 1967, Late Cretaceous Foraminiferida and paleoecology, northwest Baja California, Mexico: 180 p.
Mickey, Michael B., 1971, Upper Cretaceous biostratigraphy of a portion of northwestern Baja California, Mexico: 189 p.
Minch, John A., 1966, Stratigraphy and structure of the Tijuana-Rosarito Beach area, northwestern Baja California, Mexico: 75 p.
Moscoso, Belisario A., 1967, A thick section of Jurassic "flysch" in the Santa Ana Mountains, southern California: 106 p.

Reed, Robert G., 1967, Stratigraphy and structure of the Alisitos Formation near El Rosario, Baja California, Mexico: 118 p.
Rossetter, Robert J., 1974, Geology of the small islands of the Guaymas lineament: 92 p.
Schroeder, James E., 1967, Geology of a portion of the Ensenada quadrangle, Baja California, Mexico: 74 p.
Slyker, Robert G., 1970, Geologic and geophysical reconnaissance of the Valle de San Felipe region, Baja California, Mexico: 97 p.
Stump, Thomas E., 1972, Stratigraphy and paleontology of the Imperial Formation in the western Colorado Desert: 132 p.

Appendix 2. Senior and Other Undergraduate Research Reports

The following is a list of undergraduate research reports completed by students at San Diego State University concerning the state of Baja California. Several are bound together (U.r.r.), and the volumes do not circulate. Because of the uneven style of these reports, any given report may not photocopy well. Photocopies are available through the Interlibrary Loan Office of San Diego State University.

Acosta, Marcial G., 1963, The mineralogy and texture of the sedimentary rocks of the Rosario Formation, Punta China-Punta San Jose area, Baja California, Mexico: U.r.r., v. 7, pt. 1.
———1965, Paleontology of the late Pleistocene marine terraces, Punta Baja, Baja California, Mexico: U.r.r., v. 9, pt. 1.
Andersen, Robert L., 1966, Geology of the El Rosarito region, Baja California, Mexico: U.r.r., v. 10, pt. 1.
Ashley, Randal J., 1972, The geology of an area northeast of the Guadalupe Valley, Baja California, Mexico: U.r.r., v. 20, pt. 1.
Bell, Robert E., 1969, Geology of a plutonic-metamorphic terrain near Rancho Vallecitos, northwestern Baja California, Mexico: U.r.r., v. 14, pt. 1.
Birkhahn, Phil, 1969, Structure and petrology of a tonalite pluton in Baja California: U.r.r., v. 13.
Burk, Robert L., 1969, Petrogenetic implications of a basaltic anorthosite, Santa Clara Valley, Baja California, Mexico: U.r.r., v. 13.
Carpenter, William B., 1968, Stratigraphy of the Rosario Formation, northwestern Baja California, Mexico: U.r.r., v. 12, pt. 1.
Craig, James D., 1972, Geology of the Colnet-Camalú area, northwestern Baja California, Mexico: U.r.r., v. 19.
Feldman, Lawrence H., 1963, The identification of species and other information from the following localities: Near Descanso NW Baja California, near salt works—Imperial Beach, Bonita Woods—San Diego County; and from the Harrington Collection: U.r.r., v. 7, pt. 2.
Flynn, Michael R., 1968, Reconnaissance Bouguer gravity of Baja California, Mexico: U.r.r., v. 12, pt. 2.
Frazer, Martin, 1972, Geology of Valle de las Plamas: U.r.r., v. 20, pt. 1.
Gunther, Fred J., 1963, Ostracoda of certain Eocene rocks of San Diego: U.r.r., v. 7, pt. 3.
Helander, Peter, 1969, Cretaceous geology of an area north of Ensenada, along Hwy 1, n. w. Baja California, Mexico: U.r.r., v. 14, pt. 1.
Henry, Arthur J., 1966, Reconnaissance geology of the southeastern Sierra de Pintas, Baja California: U.r.r., v. 10, pt. 3.

Hofman, George R., 1969, Geology of La Mision, Baja California, Mexico: U.r.r., v. 13.
Holden, John C., 1961, The paleontology of some Cretaceous ostracods from the Rosario Formation: U.r.r., v. 5., pt. 2.
Hord, Peter R., 1965, Sedimentary structures of the Rosario Formation east of Rosarito Beach, Baja California, Mexico: U.r.r., v. 9, pt. 2.
Itson, Joe A., 1970, Engineering geology of coastal landslides north of Ensenada, Mexico: U.r.r., v. 16, pt. 1.
James, Allen L., 1972, Geochemical evaluation of temperature at depth for a proposed geothermal reservoir in the Ensenada-Punta Banda area, Baja California, Mexico: U.r.r., v. 20, pt. 2.
James, A. Harvey, 1966, The occurrence and origin of axinite at the La Olivia scheelite mine, Baja California, Mexico.
Jones, William, 1966, General geology of San Felipe, Baja California, Mexico: U.r.r., v. 10, pt. 3.
Kaiser, John V., Jr., 1967, Geology of Rancho Agua Caliente, Baja California, Mexico: U.r.r., v. 11, pt. 3.
Kebort, A. T., 1964, Geology of the Mesa Redonda area, northwest Baja California, Mexico: U.r.r., v. 8, pt. 3.
Kennedy, George L., 1969, The west American Cenozoic Pholadidae (Mollusca: Bivalvia): U.r.r., v. 13.
La Borde, Ronald T., 1967, Reconnaissance geology of the northern Sierra de Pintas, Baja California: U.r.r., v. 11, pt. 3.
Layson, Tim, 1963, Paleontology of the mudstone-siltstone member of the Rosario Formation, Punta San José, Baja California, Mexico: U.r.r., v. 7, pt. 3.
Leedom, Steve, 1967, A stratigraphic study of the Alisitos Formation near San Vicente, Baja California, Mexico: U.r.r., v. 11, pt. 4.
Lehrer, Lloyd, 1968, Stratigraphy of part of the Alisitos Formation, Santo Tomás Valley, Baja California, Mexico: U.r.r., v. 12, pt. 2.
Lehtola, Richard, 1969, Geology of an area of the Sierra Mayor, Baja California, Mexico: U.r.r., v. 13.
McGee, David C., 1965, Upper Cretaceous (Campanian) Foraminiferida from Punta Baja, Baja California: U.r.r., v. 9, pt. 2.
———1966, A note on the distribution of selected groups of Upper Cretaceous Foraminiferida and their value as environmental indicators: U.r.r., v. 10, pt. 3.
———1966, Upper Cretaceous Foraminiferida from Valle El Morro, Baja California, Mexico. U.r.r., v. 10, pt. 3.
———1967, Some Miocene diatoms from northwestern Baja California: U.r.r., v. 11, pt. 4.
Minch, John A., 1964, Geology and paleontology of the Playas de Tijuana area, northwestern Baja California, Mexico: U.r.r., v. 8, pt. 4.
Muehlberg, G. E., 1968, Distribution of benthonic foraminifera from the "slope-break" area along Baja California, Mexico: U.r.r., v. 12, pt. 2.
Pendarvis, Lawrence D., 1969, Geology of the El Descanso area, northwestern Baja California, Mexico: U.r.r., v. 14, pt. 2.
Peterson, Bruce A., 1967, Geologic map of northern Rancho Agua Caliente, Baja California, Mexico: U.r.r., v. 11, pt. 5.
Petrick, William, 1972, Agua Blanca geothermal investigation, resistivity survey: U.r.r., v. 20, pt. 2.
Pfister, William F., 1966, Geology of a portion of the northern Sierra Juárez, Baja California, Mexico: U.r.r., v. 10, pt. 3.
Rogers, Mark H., 1970, Geology of San Matias Pass area, Baja California: U.r.r., v. 16, pt. 1.
Rossetter, Robert J., 1970, Geology of La Encantada Island, Baja California, Mexico: U.r.r., v. 16, pt. 2.
Ryan, Douglas P., 1963, A petrographic and petrologic study of the Alisitos Formation near Punta China, Baja California, Mexico: U.r.r., v. 7, pt. 3.

Schatzinger, Richard A., 1972, Pliocene molluscan paleoecology: San Diego Formation south of the Tia Juana River: U.r.r., v. 20, pt. 3.
Schulte, Kenneth C., 1966, Geology of an area east of La Misión de San Miguel, Baja California: U.r.r., v. 10, pt. 4.
Shapley, Fred, 1963, A foraminiferal fauna from the Punta China area, Baja California, Mexico: U.r.r., v. 7, pt. 3.
Smith, Warren L., 1967, Stratigraphy of the Cerro Coronel area, northwestern Baja California, Mexico: U.r.r., v. 11, pt. 5.
Snyder, Henry M., 1970, Mesozoic dike swarm in Baja California: U.r.r., v. 16, pt. 2.
Strandstra, Robert, 1972, A self-potential survey of part of the Agua Blanca geothermal zone: U.r.r., v. 20, pt. 3.
Strong, Larry J., 1971, A study of a group of andesitic plugs in the Tijuana-Tecate area of Baja California, Mexico: U.r.r., v. 18.
Stuart, Charles J., 1964, Geology of a portion of the area east of Rosarito Beach, Baja California: U.r.r., v. 8, pt. 4.
Stump, Thomas E., 1970, Stratigraphy and paleontology of the Imperial Formation in the western Colorado Desert: U.r.r., v. 16, pt. 2.
Turner, John M., 1964, The geology of the Rosarito Beach area: U.r.r., v. 8, pt. 4.
Worthington, Bryan C., 1965, Sedimentary structures of the Rosario Formation in El Morro Valley, Baja California, Mexico: U.r.r., v. 9, pt. 2.

References Cited

Acosta, M. G., 1970, Upper Cretaceous geology of the Bahia Soledad-Punta China area, in Pacific slope geology of northern Baja California and adjacent Alta California: Am. Assoc. Petroleum Geologists and Soc. Econ. Paleontologists and Mineralogists, Pacific Secs., p. 30-36.
Aerospace Center of Defense Mapping Agency, 1969, Operational navigation chart H-22 (8th ed.): Aerospace Center of Defense Mapping Agency, scale 1:1,000,000.
Allen, C. R., 1968, The tectonic environments of seismically active and inactive areas along the San Andreas fault system, in Conference on geologic problems of San Andreas fault system, Stanford, Calif., 1967, Proc.: Stanford Univ. Pubs. Geol. Sci., v. 11, p. 70-80.
Allen, C. R., Silver, L. T., and Stehli, F. G., 1960, Agua Blanca fault—A major transverse structure of northern Baja California, Mexico: Geol. Soc. America Bull., v. 71, p. 457-482.
Allen, C. R., Allison, E. C., Roberts, E. R., and Silver, L. T., 1961, Geology of northwestern Baja California, Mexico, in Geol. Soc. America, Cordilleran Sec., Guidebook for field trips, San Diego, 1961: San Diego, Calif., San Diego State College, p. 55-65.
Allison, E. C., 1955, Middle Cretaceous gastropoda from Punta China, Baja California, Mexico: Jour. Paleontology, v. 29, p. 400-432.
―――1964, Geology of areas bordering Gulf of California, in van Andel, Tj. H., and Shor, G. G., Jr., eds., Marine geology of the Gulf of California—A symposium: Am. Assoc. Petroleum Geologists Mem. 3, p. 3-29.
―――1969, Mesozoic bivalvia of the Peninsular Ranges: Baja California and southern California [abs.]: Geol. Soc. America Spec. Paper 121, p. 3-4.
Alonso Espinosa, H., and Mooser, F., 1964, El pozo M-3 del campo geotérmico del Cerro Prieto, B.C., México: Asoc. Mexicana Geólogos Petroleros Bol., v. 16, p. 163-177.
Anda, see de Anda.
Andersen, R. L., 1969, Geology of the Playa San Felipe 30 × 40 minute quadrangle, Baja California, Mexico [abs.]: Geol. Soc. America Spec. Paper 121, p. 479.
Anderson, C. A., 1950, Geology of islands and neighboring land areas, Pt. 1 of 1940 E. W. Scripps cruise to the Gulf of California: Geol. Soc. America Mem. 43, 53 p.

References Cited

Anderson, F. M., and Hanna, G. D., 1935, Cretaceous geology of Lower California: California Acad. Sci. Proc., ser. 4, v. 23, p. 1-34.

Armstrong, R. L., and Suppe, J., 1973, Potassium-argon geochronometry of Mesozoic igneous rocks in Nevada, Utah, and southern California: Geol. Soc. America Bull., v. 84, p. 1375-1392.

Arnold, B. A., 1957, Late Pleistocene and recent changes in land forms, climate, and archaeology in central Baja California: California Univ. Pubs. Geography, v. 10, p. 201-318.

Arnold, R., 1906, Tertiary and Quaternary pectens of California: U.S. Geol. Survey Prof. Paper 47, 28 p.

Artim, E. R., and Pinckney, C. J., 1973, La Nacion fault system, San Diego, California: Geol. Soc. America Bull., v. 84, p. 1075-1080.

Aschmann, H., 1959, The central desert of Baja California—Demography and ecology: California Univ. Press, 315 p.

Atwater, T., 1970, Implications of plate tectonics for the Cenozoic tectonic evolution of western North America: Geol. Soc. America Bull., v. 81, p. 3513-3536.

———1971, Evidence from plate tectonics for the age of initiation of deformation on the San Andreas-Gulf of California system: Geol. Soc. America Abs. with Programs, v. 3, p. 75-76.

Automobile Club of Southern California, 1973, Baja California norte: Los Angeles, Automobile Club of Southern California, 80 p.

Baegert, J. J., 1771, Observations in Lower California (translated by Brandenburg, M. M., and Baumann, C. L., 1952): California Univ. Press, 218 p.

Bailey, E. H., Irwin, W. P., and Jones, D. L., 1964, Franciscan and related rocks, and their significance in the geology of western California: California Div. Mines and Geology Bull. 183, 177 p.

Banks, P. O., and Silver, L. T., 1969, U-Pb isotope analyses of zircons from Cretaceous plutons of the Peninsular and Transverse Ranges, southern California [abs.]: Geol. Soc. America Spec. Paper 121, p. 17-18.

Barnard, F. L., 1968a, Structure and tectonics of the Sierra Cucapas, northeastern Baja California—Progress report [abs.]: Geol. Soc. America Spec. Paper 115, p. 311-312.

———1968b, Structural geology of the Sierra de los Cucapas, northeastern Baja California, Mexico, and Imperial County, California [Ph.D. dissert.]: Boulder, Univ. Colorado, 157 p. (Univ. Microfilms, Inc., no. 69-4326).

Beal, C. H., 1948, Reconnaissance of the geology and oil possibilities of Baja California, Mexico: Geol. Soc. America Mem. 31, 138 p.

Bellemin, G. J., and Merriam, R. H., 1958, Petrology and origin of the Poway Conglomerate, San Diego County, California: Geol. Soc. America Bull., v. 69, p. 199-220.

Biehler, S., Kovach, R. L., and Allen, C. R., 1964, Geophysical framework of northern end of Gulf of California structural province, in van Andel, Tj. H., and Shor, G. G., Jr., eds., Marine geology of the Gulf of California—A symposium: Am. Assoc. Petroleum Geologists Mem. 3, p. 126-143.

Blake, W. P., 1858, Report of a geological reconnaissance in California: New York, H. Baillicre, p. 104-108, 122-129.

Bonillas, Y. S., and Urbina, F., 1913, Informe acerca de los recursos naturales de la parte norte de la Baja California, especialmente del Delta del Río Colorado: Inst. Geológico de México, Parergones IV, p. 161-235.

Böse, E., and Wittich, E., 1913, Informe relativo á la exploración de la región norte de la costa occidental de la Baja California: Inst. Geológico de México, Parergones IV, p. 307-533.

Bramkamp, R. A., 1935, Molluscan fauna of Imperial Formation of San Gorgonio Pass [abs.]: Pan-Am. Geologist, v. 62, p. 70-71; also Geol. Soc. America Proc. for 1934, p. 385.

Brooks, B., and Roberts, E., 1954, Geology of the Jacumba area, San Diego and Imperial Counties, map sheet no. 23, in Jahns, R. H., ed., Geology of southern California: California Div. Mines Bull. 170, scale 1:62,500.

Browne, J. R., 1868, Report on the mineral resources of the states and territories west

of the Rocky Mountains: Washington, D.C., U.S. Treasury Dept., 674 p.
——1869, Resources of the Pacific slope: New York, D. Appleton and Co., 200 p.
Bushee, J., Holden, J., Geyer, B., and Gastil, R. G., 1963, Lead-alpha dates for some basement rocks of southwestern California: Geol. Soc. America Bull., v. 74, p. 803-806.
Byrne, J. V., and Emery, K. O., 1960, Sediments of the Gulf of California: Geol. Soc. America Bull., v. 71, p. 983-1010.
Carey, S. W., 1958, The tectonic approach to continental drift, in Continental drift—A symposium: Hobart, Univ. Tasmania, 355 p.
Chapman, R. H., 1966, Gravity base station network: California Div. Mines and Geology Spec. Rept. 90, 49 p.
Cohen, L. H., Condie, K. C., Kuest, L. J., Jr., MacKenzie, G. S., Meister, F. H., Pushkar, P., and Stueber, A. M., 1963, Geology of the San Benito Islands, Baja California, Mexico: Geol. Soc. America Bull., v. 74, p. 1355-1370.
Combs, J., 1971, Heat flow and geothermal resource estimates for the Imperial Valley, in Rex, R. W., and others, Cooperative geological-geophysical-geochemical investigations of geothermal resources in the Imperial Valley area of California: Riverside, Univ. California, p. 5-28.
Comisión Intersecretarial Coordinadora del Levantamiento de la Cartográfica de la República Mexicana, 1958, Ensenada, Cedros, and Isla Tiburón 1:500,000 topographic quadrangles: México, D.F.
Consejo de Recursos Naturales no Renovables, 1965, Proyecto de exploración de yacimientos de minerales metálicos de la zona "Ensenda, B.C.": México, D.F., México, Consejo de Recursos Naturales no Renovables, p. 13.
Cooper, W. S., 1967, Coastal dunes of California: Geol. Soc. America Mem. 104, 131 p.
Cross, A. T., 1966, Source and distribution of palynomorphs in bottom sediments, southern part of the Gulf of California: Marine Geology, v. 4, p. 467-524.
Crowell, J. C., 1962, Displacement along the San Andreas fault, California: Geol. Soc. America Spec. Paper 71, 61 p.
Curtis, G. H., Evernden, J. F., and Lipson, J. I., 1958, Age determination of some granitic rocks in California by the potassium-argon method: California Div. Mines Spec. Rept. 54, 16 p.
Daley, R. A., 1933, The depths of the earth: Geol. Soc. America Bull., v. 44, p. 243-264.
Dall, W. H., 1898, A table of the North American Tertiary formations, correlated with one another and with those of western Europe, with annotations: U.S. Geol. Survey Ann. Rept. 18, pt. 2, p. 323-348.
Darton, N. H., 1921, Geologic reconnaissance in Baja California: Jour. Geology, v. 29, p. 720-748.
Davidson, G., 1897, The submerged valleys of the coast of California, U.S.A., and of Lower California, Mexico: California Acad. Sci. Proc., ser. 3: Geology, v. 1, p. 73-103.
de Anda, L. F., and Paredes, E., 1964, La falla de San Jacinto y su influencia sobre la actividad geotérmica en el Valle de Mexicali, B.C., México: Asoc. Mexicana Geólogos Petroleros Bol., v. 16, p. 178-181.
de Anda, L. F., Isita S., J., and Ruiz E., J., 1964, Geothermal energy, in United Nations conference on new sources of energy, Rome, 1961, Proc., Vol. 2: New York, United Nations, Geothermal Energy, pt. 1, p. 149-165.
DeLisle, M., Morgan, J. R., Heldenbrand, J., and Gastil, R. G., 1965, Lead-alpha ages and possible sources of metavolcanic rock clasts in the Poway Conglomerate, southwest California: Geol. Soc. America Bull., v. 76, p. 1069-1074.
Dibblee, T. W., Jr., 1954, Geology of the Imperial Valley region, California, in Jahns, R. H., ed., Geology of southern California: California Div. Mines Bull. 170, chap. 2, contr. 2, p. 21-28.
Dickerson, R. E., 1918, Mollusca of the Corrizo Creek beds and their Caribbean affinities [abs.]: Geol. Soc. America Bull., v. 29, p. 148.
Dickinson, W. R., and Hatherton, T., 1967, Andesitic volcanism and seismicity around the Pacific: Science, v. 157, p. 801-803.

Doe, B. R., and Tilling, R. I., 1966, The distribution of lead between coexisting K-feldspar and plagioclase [abs.]: Am. Geophys. Union Trans., v. 47, p. 205-206.

Donnelly, M., 1935, Geology and mineral deposits of the Julian district, San Diego County, California: California Jour. Mines and Geology, v. 30, p. 331-370.

Downs, T., and White, J. A., 1968, A vertebrate faunal succession in superposed sediments from late Pliocene to middle Pleistocene in California: Internat. Geol. Cong., 23rd, Prague 1968, Tertiary-Quaternary boundary, Academia, sec. 10, p. 41-47.

Dudley, P. H., 1935, Geology of a portion of the Perris block, southern California: California Jour. Mines and Geology, v. 31, p. 487-506.

Duffield, W. A., 1968, The petrology and structure of the El Pinal tonalite, Baja California, Mexico: Geol. Soc. America Bull., v. 79, p. 1351-1374.

Durham, J. W., 1950, Megascopic paleontology and marine stratigraphy, Pt. 2 of 1940 E. W. *Scripps* cruise to the Gulf of California: Geol. Soc. America Mem. 43, 216 p.

Durham, J. W., and Allison, E. C., 1960, The geologic history of Baja California and its marine faunas, in The biogeography of Baja California and adjacent seas, Pt. 1, Geologic history: Systematic Zoology, v. 9, p. 47-91.

Dutcher, L. C., Hardt, W. F., and Moyle, W. R., Jr., 1972, Preliminary appraisal of ground water in storage with reference to geothermal resources in the Imperial Valley area, California: U.S. Geol. Survey Circ. 649, 57 p.

D'Vincent, S., 1967, Primitive Sequoia not previously identified: California Garden, v. 58, p. 14-15.

Ehlig, P. L., 1971, Origin of the San Gabriel Mountains as a transverse range: Geol. Soc. America Abs. with Programs, v. 3, p. 115-116.

Elders, W. A., Rex, R. W., Meidav, T., Robinson, P. T., and Biehler, S., 1971, A plate tectonic model for the Salton Trough: Geol. Soc. America Abs. with Programs, v. 3, p. 116-117.

Elliott, W. J., 1970, Gravity survey and regional geology of the San Diego embayment, southwest San Diego County, California, in Pacific slope geology of northern Baja California and adjacent Alta California: Am. Assoc. Petroleum Geologists and Soc. Econ. Paleontologists and Mineralogists, Pacific Secs., p. 10-22.

Ellis, A. J., 1919, Physiography, in Ellis, A. J., and Lee, C. H., eds., Geology and ground waters of the western part of San Diego County, California: U.S. Geol. Survey Water-Supply Paper 446, p. 20-50.

Emerson, W. K., 1956, Pleistocene invertebrates from Punta China, Baja California, Mexico: Am. Mus. Nat. History Bull., v. 111, p. 313-342.

——1960, Pleistocene invertebrates from near Punta San Jose, Baja California, Mexico: Am. Mus. Nat. History Novitates, no. 2002, 7 p.

Emerson, W. K., and Addicott, W. O., 1958, Pleistocene invertebrates from Punta Baja, Baja California, Mexico: Am. Mus. Nat. History Novitates, no. 1909, 11 p.

Emerson, W. K., and Hertlein, L. G., 1960, Pliocene and Pleistocene invertebrates from Punta Rosalia, Baja California, Mexico: Am. Mus. Nat. History Novitates, no. 2004, 8 p.

——1964, Invertebrate megafossils of the Belvedere expedition to the Gulf of California: San Diego Soc. Nat. History Trans., v. 13, p. 333-368.

Emery, K. O., 1957, Sediments of three bays of Baja California—Sebastian Viscaino, San Cristobal, and Todos Santos: Jour. Sed. Petrology, v. 27, p. 95-115.

——1960, The sea off southern California; a modern habitat of petroleum: New York, John Wiley & Sons, Inc., 366 p.

Emery, K. O., Butcher, W. S., Gould, H. R., and Shepard, F. P., 1952, Submarine geology off San Diego, California: Jour. Geology, v. 60, p. 511-548.

Emmons, S. F., and Merrill, G. P., 1894, Geological sketch of Lower California: Geol. Soc. America Bull., v. 5, p. 489-514.

Estavillo, W., and Rogers, M., 1970, Potassium-argon dating of metamorphic and plutonic igneous rocks from San Matias Pass, Baja California, Mexico: Geol. Soc. America Abs. with Programs, v. 2, pt. 2, p. 90-91.

Everhart, D. L., 1951, Geology of the Cuyamaca Peak quadrangle, California: California Div. Mines Bull. 159, p. 51-115.

Evernden, J. F., and Kistler, R. W., 1970, Chronology of emplacement of Mesozoic batholithic complexes in California and western Nevada: U.S. Geol. Survey Prof. Paper 623, 42 p.

Facca, G., 1966, La energía geotérmica: Rev. Mexicana Electricidad, p. 5-12, 35-38.

Fairbanks, H. W., 1893, Geology of San Diego County, also of portions of Orange and San Bernardino Counties: California State Mining Bur. Ann. Rept. 11, p. 76-120.

Fife, D. L., 1969, Reconnaissance geology of the Bahia Santa Rosalia quadrangle, Baja California, Mexico [abs.]: Geol. Soc. America Spec. Paper 121, p. 505-506.

Fife, D. L., Minch, J. A., and Crampton, P. J., 1967, Late Jurassic age of the Santiago Peak Volcanics, California: Geol. Soc. America Bull., v. 78, p. 299-304.

Fisher, R. L., and Hess, H. H., 1963, Trenches, in The sea—Ideas and observations in progress, Vol. 3, The earth beneath the sea (and) history: New York, Interscience Pubs., p. 411-436.

Flores, T., 1931, Explicación de la carta geológica de la Baja California: Inst. Geológico de México, Cartas Geológicas y Mineras, no. 1, 22 p.

Flores, T., and González, P., Jr., 1913, Exploración de la parte central elevada de la porción norte de la Península de la Baja California: Inst. Geológico de México, Parergones IV, p. 237-275.

Flynn, C. J., 1970, Post-batholithic geology of the La Gloria-Presa Rodríguez area, Baja California, Mexico: Geol. Soc. America Bull., v. 81, p. 1789-1806.

Fries, C., Jr., and Schmitter, E., 1945, Scheelite deposits in the northern part of the Sierra de Juarez, Northern Territory, Lower California, Mexico: U.S. Geol. Survey Bull. 946-C, pt. IV, p. 73-101; México Com. Dir. Inv. Rec. Miner. Bol. 2 (Spanish translation).

Fuchs, E.P.J., 1886, Note sur les gisements de cuivre du Boléo (Lower California): Acad. Sci. Comptes Rendus, v. 14, pt. 2, p. 410-426.

Gabb, W. M., 1882, Notes on the geology of Lower California: California Geol. Survey, pt. 2: Geology, app., p. 137-148.

Gastil, R. G., 1961, The elevated erosion surfaces, in Geol. Soc. America, Cordilleran Sec., Guidebook for field trips, San Diego, 1961: San Diego, Calif., San Diego State College, p. 1-4, map.

Gastil, R. G., and Allison, E. C., 1966, An Upper Cretaceous fault-line coast [abs.]: Am. Assoc. Petroleum Geologists Bull., v. 50, p. 647-648.

Gastil, R. G., and Krummenacher, D., 1970, Reconnaissance potassium-argon dates for Cenozoic volcanic rocks in the state of Baja California: Geol. Soc. America Abs. with Programs, v. 2, p. 92-93.

Gastil, R. G., Stickney, D., and Terry, A., 1971, Pluton sizes in the Peninsular Ranges of Baja California and the Sierra Nevada: Geol. Soc. America Abs. with Programs, v. 3, p. 123.

Gastil, R. G., Phillips, R. P., and Rodríguez-Torres, R., 1972, The reconstruction of Mesozoic California: Internat. Geol. Cong., 24th, Montreal 1972, sec. 3, p. 217-229.

Gastil, R. G., LeMone, D. V., and Stewart, W. J., 1973, Permian fusulinids from near San Felipe, Baja California: Am. Assoc. Petroleum Geologists Bull., v. 57, p. 746-747.

Gerhard, P., and Gulick, H. E., 1970, Lower California guidebook (4th ed.): Glendale, California, Arthur H. Clark Company, 212 p.

Gómez, M., 1971, Sobre la presencia de estratos marinos del Mioceno en el Estado de Sonora, México: Rev. Inst. Mexicano Petróleo, v. 3, no. 4, p. 77-78.

Goodyear, W. A., 1888, San Diego County, in Mineral resources of the state: California State Mining Bur., Ann. Rept. 8, p. 512-528, 694.

Gorsline, D. S., and Stewart, R. A., 1962, Benthic marine exploration of Bahia de San Quintin, Baja California, 1960-1961, marine and Quaternary geology: Pacific Naturalist, v. 3, p. 282-319.

Grant, U. S., IV, and Gale, H. R., 1931, Catalogue of the marine Pliocene and Pleistocene mollusca of California: San Diego Soc. Nat. History Mem. 1, pt. 1, 1036 p.

Grewingk, C., 1848, Beitrag zur Kenntniss der geognostischen Beschaffenheit Californiens:

St. Petersburg, Russian K. Min. Ges., Verh. 1847, p. 142-162.
Hamilton, W., 1961, Origin of the Gulf of California: Geol. Soc. America Bull., v. 72, p. 1307-1318.
——1969a, The volcanic central Andes—A modern model for the Cretaceous batholiths and tectonics of western North America, in Andesite Conf., Eugene and Bend, Oreg., 1968, Proc. (Internat. Upper Mantle Proj. Sci. Rept. 16): Oregon Dept. Geology and Mineral Industries Bull. 65, p. 175-184.
——1969b, Mesozoic California and the underflow of Pacific mantle: Geol. Soc. America Bull., v. 80, p. 2409-2430.
——1971, Recognition on space photographs of structural elements of Baja California: U.S. Geol. Survey Prof. Paper 718, 26 p.
Hanna, G. D., 1925, Expedition to Guadalupe Island, Mexico, in 1922, general report: California Acad. Sci. Proc., ser. 4, v. 14, p. 217-275.
——1926, Paleontology of Coyote Mountain, Imperial County, California: California Acad. Sci. Proc., ser. 4, v. 14, p. 427-503.
——1927, Geology of the west Mexican islands: Pan-Am. Geologist, v. 48, no. 1, p. 1-24.
Hanna, G. D., and Hertlein, L. G., 1927, Expedition of California Academy of Science to the Gulf of California in 1921—Geology and paleontology: California Acad. Sci. Proc., ser. 4, v. 16, p. 137-157.
Hanna, M. A., 1926, Geology of the La Jolla quadrangle, California: California Univ. Dept. Geol. Bull., v. 16, p. 187-246.
Harrison, J. C., and Mathur, S. P., 1964, Gravity anomalies in Gulf of California, in van Andel, Tj. H., and Shor, G. G., Jr., eds., Marine geology of the Gulf of California—A symposium: Am. Assoc. Petroleum Geologists Mem. 3, p. 76-89.
Hawkins, J. W., 1970a, Petrology and possible tectonic significance of late Cenozoic volcanic rocks, southern California and Baja California: Geol. Soc. America Bull., v. 81, p. 3323-3338.
——1970b, Metamorphosed late Jurassic andesites and dacites of the Tijuana-Tecate area, Baja California, in Pacific slope geology of northern Baja California and adjacent Alta California: Am. Assoc. Petroleum Geologists and Soc. Econ. Paleontologists and Mineralogists, Pacific Secs., p. 25-29.
Heim, A., 1915, Sur la géologie de la partie méridionale de la Basse California: Acad. Sci. Comptes Rendus 161, p. 419-422.
——1921, Vulkane in der Ungebung der Oase la Purisima auf der Halbinsel Nieder Kalifornia: Zeitschr. Vulkanologie, Bd. 6, H. 1, p. 15-21.
——1922, The Tertiary of southern Lower California: Geol. Mag., v. 59, p. 529-547.
Henyey, T. L., 1971, Heat flow in the northern Gulf of California: Geol. Soc. America Abs. with Programs, v. 3, p. 135.
Henyey, T. L., and Bischoff, J. L., 1973, Tectonic elements of the northern part of the Gulf of California: Geol. Soc. America Bull., v. 84, p. 315-330.
Hertlein, L. G., 1925, Pectens from the Tertiary of Lower California: California Acad. Sci. Proc., ser. 4, v. 14, p. 1-35.
Hertlein, L. G., and Grant, U. S., IV, 1939, Geology and oil possibilities of southwestern San Diego County: California Jour. Mines and Geology, v. 35, p. 57-78.
——1944, The geology and paleontology of the marine Pliocene of San Diego, California: Pt. 1: Geology: San Diego Soc. Nat. History Mem. 2, p. 1-72.
Hertlein, L. G., and Jordan, E. K., 1927, Paleontology of the Miocene of Lower California: California Acad. Sci. Proc., ser. 4, v. 16, p. 605-646.
Hilde, T.W.C., 1964, Magnetic profiles across Gulf of California, in van Andel, Tj. H., and Shor, G. G., Jr., eds., Marine geology of Gulf of California—A symposium: Am. Assoc. Petroleum Geologists Mem. 3, p. 122-125.
Hill, M. L., and Dibblee, T. W., Jr., 1953, San Andreas, Garlock, and Big Pine faults, California—A study of the character, history, and tectonic significance of their displacements: Geol. Soc. America Bull., v. 64, p. 443-458.
Hirschi, H., 1926, Beiträge zur Petrographie von Baja California, Mexiko: Schweizer.

Mineralog. u. Petrog. Mitt., v. 6, p. 346-350.
Hirschi, H., and de Quervain, F., 1927-1933, Beiträge zur Petrographie von Baja California: Schweizer. Mineralog. u. Petrog. Mitt., v. 7 [1927], p. 142-164; v. 8 [1928], p. 323-326; v. 10 [1930], p. 228-272; v. 13 [1933], p. 232-277.
Hodgson, R. M., 1959, The null vector as a guide to regional tectonic patterns, in The dynamics of faulting with special reference to the fault-plane work—A symposium: Canada Dominion Observatory Pub., v. 20, p. 369-384.
Holden, J. C., 1964, Upper Cretaceous ostracods from California: Paleontology, v. 7, p. 393-429.
Hudson, F. S., 1922, Geology of the Cuyamaca region of California with special reference to the origin of the nickeliferous pyrrhotite: California Univ. Dept. Geology Sci. Bull., v. 13, p. 175-252.
Imlay, R. W., 1963, Jurassic fossils from southern California: Jour. Paleontology, v. 37, p. 97-107.
——1964, Middle and Upper Jurassic fossils from southern California: Jour. Paleontology, v. 38, p. 505-509.
Ingle, J. C., Jr., 1973, Summary comments on Neogene biostratigraphy, physical stratigraphy, and paleo-oceanography in the marginal northeastern Pacific Ocean: Initial Repts. Deep Sea Drilling Project, v. 18, p. 949-960.
——1974, Paleobathymetric history of Neogene marine sediments, Northern Gulf of California, in Geology of peninsular California: Am. Assoc. Petroleum Geologists, Soc. Econ. Paleontologists and Mineralogists, Soc. Exploration Geophysicists, Pacific Secs., Guidebook for fieldtrip, p. 121-138.
Inman, D. L., Ewing, G. C., and Corliss, J. B., 1966, Coastal sand dunes of Guerrero Negro, Baja California, Mexico: Geol. Soc. America Bull., v. 77, p. 787-802.
Ives, R. L., 1962, Dating of the 1746 eruption of Tres Virgenes volcano, Baja California del Sur, Mexico: Geol. Soc. America Bull., v. 73, p. 647-648.
Johnson, I. M., 1924, Expedition of the California Academy of Sciences to the Gulf of California in 1921—The botany (the vascular plants): California Acad. Sci. Proc., ser. 4, v. 12, p. 951-1218.
Jordan, E. K., 1926, Molluscan fauna of the Pleistocene of San Quintin Bay, Lower California: California Acad. Sci. Proc., ser. 4, v. 15, p. 241-255.
Karig, D. E., and Jensky, W., 1972, The proto-Gulf of California: Earth and Planetary Sci. Letters, v. 17, p. 169-174.
Kennedy, M. P., and Moore, G. W., 1971, Stratigraphic relations of Upper Cretaceous and Eocene formations, San Diego coastal area, California: Am. Assoc. Petroleum Geologists Bull., v. 55, p. 709-722.
Kew, W. S., 1914, Tertiary echinoids of the Corrizo Creek region in the Colorado Desert: California Univ. Dept. Geology Bull., v. 8, p. 39-60.
Kilmer, F. H., 1963, Cretaceous and Cenozoic stratigraphy and paleontology, El Rosario area, Baja California, Mexico [Dissert.]: Berkeley, California Univ., Berkeley, 216 p. (Univ. Microfilms, Inc., no. 66-15526).
——1965, Late Cretaceous stratigraphy and paleontology, El Rosario, northern Baja California, Mexico [abs.]: Geol. Soc. America, Cordilleran Sec. Program, p. 31-32.
——1969, Preliminary report on the geology of Cedros Island, Baja California, Mexico [abs.]: Geol. Soc. America Spec. Paper 121, p. 521.
King, R. E., 1939, Geological reconnaissance in northern Sierra Madre Occidental of Mexico: Geol. Soc. America Bull., v. 50, p. 1625-1722.
Kirk, M. V., and MacIntyre, J. R., 1951, Cretaceous deposits of the Punta San Isidro area, Baja California [abs.]: Geol. Soc. America Bull., v. 62, p. 1505.
Kovach, R. L., and Monges C., J., 1961, Medidas de gravedad en la parte norte de Baja California: México Univ. Nac. Autónoma Inst. Geofísica Anales, v. 7, p. 9-14.
Kovach, R. L., Allen, C. R., and Press, F., 1962, Geophysical investigations in the Colorado Delta region: Jour. Geophys. Research, v. 67, p. 2845-2871.
Krause, D. C., 1965, Tectonics, bathymetry, and geomagnetism of the southern continental

borderland west of Baja California, Mexico: Geol. Soc. America Bull., v. 76, p. 617-650.

Krummenacher, D., Gastil, R. G., Bushee, J., and Doupont, J., 1975, K-Ar apparent ages, Peninsular Ranges batholith, southern California and Baja California: Geol. Soc. America Bull., v. 86, no. 6, p. 760-768.

Kuno, H., 1969, Andesites in time and space, in Andesite Conf., Eugene and Bend, Oreg., 1968, Proc. (Internat. Upper Mantle Proj., Sci. Rept. 16): Oregon Dept. Geology and Mineral Industries Bull. 65, p. 13-20.

Lamb, T. N., 1970, Fossiliferous Triassic(?) meta-sedimentary rocks near Sun City, Riverside County, California: Geol. Soc. America Abs. with Programs, v. 2, p. 110-111.

Larsen, E. S., Jr., 1948, Batholith and associated rocks of Corona, Elsinore, and San Luis Rey quadrangles, southern California: Geol. Soc. America Mem. 29, 182 p.

Larsen, E. S., Jr., Gottfried, D., Jaffe, H. W., and Waring, C. L., 1958, Lead-alpha ages of the Mesozoic batholiths of western North America: U.S. Geol. Survey Bull. 1070-B, p. 35-62.

Larson, R. L., Menard, H. W., and Smith, S. M., 1968, Gulf of California—A result of ocean-floor spreading and transform faulting: Science, v. 161, p. 781-784.

Lawson, A. C., 1893, Post-Pliocene diastrophism of the coast of southern California: California Univ. Dept. Geology Bull., v. 1, p. 115-160.

Lee, W.H.K., and Uyeda, S., 1965, Review of heat flow data, in Terrestrial heat flow: Am. Geophys. Union Geophys. Mon. 8, p. 87-190.

Leeman, W. P., and Rogers, J. J., 1970, Late Cenozoic alkali-olivine basalts of the Basin-Range province, USA: Contr. Mineralogy and Petrology, v. 25, p. 1-24.

Leeman, W. P., Gastil, R. G., and Krummenacher, D., 1972, Strontium isotopic composition of Baja California lavas [abs.]: EOS (Am. Geophys. Union Trans.), v. 53, p. 277.

Lindgren, W., 1888, Notes on the geology of Baja California, Mexico: California Acad. Sci. Proc., ser. 2, v. 1, p. 173-196.

——1889, Petrographical notes from Baja California, Mexico: California Acad. Sci. Proc., ser. 2, v. 2, p. 1-17.

——1890, Notes on the geology and petrography of Baja California, Mexico: California Acad. Sci. Proc., ser. 2, v. 3, p. 25-33.

Longinos Martínez, J., 1792, California in 1792, the expedition of José Longinos Martínez (translated by Simpson, 1938): San Marino, California, Huntington Lib. Pubs., 111 p.

Lowman, P. D., 1972, The third planet: Zurich, Weltflugbild, 170 p.

Mandra, Y. T., and Mandra, H., 1972, Paleoecology and taxonomy of silicoflagellates from an upper Miocene diatomite near San Felipe, Baja California, Mexico: California Acad. Sci. Occasional Paper 99, 35 p.

Marland Oil Company of Mexico, 1924, Informe sobre la explorción geológica de la Baja California: Bol. Petróleo, v. 17, no. 6, p. 417-453; v. 18, no. 1, p. 14-53.

Martínez, P. L., 1960, A history of Lower California (1st English ed.): México, D.F., Priv. Pub., 567 p.

Maytum, J. R., and Elliott, W. J., 1970, Upper Cretaceous strata of the La Jolla-Point Loma area, San Diego: Correlation and physical stratigraphy, in Pacific slope geology of northern Baja California and adjacent Alta California: Am. Assoc. Petroleum Geologists and Soc. Econ. Paleontologists and Mineralogists, Pacific Secs., p. 38-52.

McEldowney, R. C., 1970, An occurrence of Paleozoic fossils in Baja California, Mexico: Geol. Soc. America Abs. with Programs, v. 2, p. 117.

McFall, C. C., 1968, Reconnaissance geology of the Concepcion Bay area, Baja California, Mexico: Stanford Univ. Pubs. Geol. Sci., v. 10, no. 5, 25 p.

McGee, D. C., 1965, Upper Cretaceous (Campanian) Foraminiferida from Punta Baja, Baja California [abs.]: Am. Assoc. Petroleum Geologists Bull., v. 49, p. 1087.

Meigs, P., III, 1935, The Dominican mission frontier of Lower California: California Univ. Pubs. Geography, v. 7, pt. VI, p. 1-232.

Menard, H. W., Jr., 1960, The East Pacific Rise: Science, v. 132, p. 1737-1746.

Menard, H. W., Jr., and Atwater, T., 1968, Changes in direction of sea floor spreading: Nature, v. 219, no. 5153, p. 463-467.

Mercado, S., 1969, Cerro Prieto geothermal field, Baja California, Mexico [abs.]: EOS (Am. Geophys. Union Trans.), v. 50, p. 59.

Merifield, P. M., Lamar, D. L., and Stout, M. L., 1971, Geology of central San Clemente Island, California: Geol. Soc. America Bull., v. 82, p. 1989-1994.

Merriam, R. H., 1946, Igneous and metamorphic rocks of the southwestern part of the Ramona quadrangle, San Diego, California: Geol. Soc. America Bull., v. 57, p. 223-260.

―― 1959, Geology of the Santa Ysabel quadrangle, San Diego County, California: California Div. Mines Bull. 177, p. 7-20.

―― 1965, San Jacinto fault in northwestern Sonora, Mexico: Geol. Soc. America Bull., v. 76, p. 1051-1054.

Merrill, F.J.H., 1914, Geology and mineral resources of San Diego and Imperial Counties: San Francisco, California State Mining Bur., 113 p.

Merrill, G. P., 1897, Notes on the geology and natural history of the peninsula of Lower California: U.S. Natl. Mus. Rept. for 1895, p. 969-994.

Miller, F. S., 1937, Petrology of the San Marcos gabbro, southern California: Geol. Soc. America Bull., v. 48, p. 1397-1426.

Miller, W. J., 1935a, A geologic section across the southern Peninsular Ranges of California: California Jour. Mines and Geology, v. 31, p. 115-142.

―― 1935b, Geomorphology of the southern Peninsular Range of California: Geol. Soc. America Bull., v. 46, p. 1535-1562.

―― 1946, Crystalline rocks of southern California: Geol. Soc. America Bull., v. 57, p. 457-540.

Milow, E. D., and Ennis, D. B., 1961, Guide to geologic field trip no. 2, southwestern San Diego County, in Geol. Soc. America, Cordilleran Sec., Guidebook for field trips, San Diego, 1961: San Diego, Calif., San Diego State College, p. 23-43.

Mina Uhink, F., 1956, Bosquejo geológico de la parte sur de la Península de Baja California: Internat. Geol. Cong., 20th, Mexico [D.F.] 1956, Excursion A-7, p. 11-47.

―― 1957, Bosquejo geológico del Territorio Sur de la Baja California: Asoc. Mexicana Geólogos Petroleros Bol., v. 9, p. 139-269.

Minch, J. A., 1967, Stratigraphy and structure of the Tijuana-Rosarito Beach area, northwestern Baja California, Mexico: Geol. Soc. America Bull., v. 78, p. 1155-1178.

―― 1969, A depositional contact between the pre-batholithic Jurassic and Cretaceous rocks in Baja California, Mexico: Geol. Soc. America Abs. with Programs for 1969, pt. 3, p. 42-43.

―― 1970, Early Tertiary paleogeography of a portion of the northern Peninsular Ranges, in Pacific slope geology of northern Baja California and adjacent Alta California: Am. Assoc. Petroleum Geologists and Soc. Econ. Paleontologists and Mineralogists, Pacific Secs., p. 83-87.

Minch, J. A., Schulte, K. C., and Hofman, G., 1970, A middle Miocene age for the Rosarito Beach Formation in northwestern Baja California, Mexico: Geol. Soc. America Bull., v. 81, p. 3149-3154.

Miyashiro, A., 1962, Evolution of metamorphic belts: Jour. Petrology, v. 2, p. 277-311.

Mohr, P. A., 1970, Catalog of chemical analysis of rocks from the intersection of the African, Gulf of Aden, and Red Sea rift systems: Smithsonian Contr. Earth Sci., no. 2, 262 p.

Moore, D. G., 1969, Reflection profiling studies of the California continental borderland: Structure and Quaternary turbidite basins: Geol. Soc. America Spec. Paper 107, 142 p.

―― 1973, Plate-edge deformations and crustal growth, Gulf of California structural province: Geol. Soc. America Bull., v. 84, p. 1883-1906.

Moore, D. G., and Buffington, E. C., 1968, Transform faulting and growth of the Gulf of California since the late Pliocene: Science, v. 161, no. 3847, p. 1238-1241.

Moore, G. W., 1972, Offshore extension of the Rose Canyon fault, San Diego, California: U.S. Geol. Survey Prof. Paper 800-C, p. C113-C116.

Moore, J. G., 1959, The quartz diorite boundary line in the western United States: Jour. Geology, v. 67, p. 198-210.

Mooser, F., 1964, Provincias geotérmicas de México: Asoc. Mexicana Geólogos Petroleros Bol., v. 16, p. 153-161.

Mooser, F., and Rayes L., A., 1961, El grupo volcánico de Las Tres Vírgenes, Mpio. de Santa Rosalía, Territorio de Baja California, México: México Univ. Nac. Autónoma Inst. Geología Bol. 61, p. 47-48.

Morris, W. J., 1966, Fossil mammals from Baja California, new evidence on early Tertiary migrations: Science, v. 153, p. 1376-1378.

―――1967, Baja California—Late Cretaceous dinosaurs: Science, v. 155, p. 1539-1541.

Morton, P. K., ed., 1972, Geologic guidebook to the northern Peninsular Ranges, Orange and Riverside Counties, California: Natl. Assoc. Geology Teachers, Far Western Sec., and South Coast Geol. Soc., Chapman College, Orange, California, 96 p.

Muffler, L.J.P., and White, D. E., 1969, Active metamorphism of upper Cenozoic sediments in the Salton Sea geothermal field and the Salton Trough, southeastern California: Geol. Soc. America Bull., v. 80, p. 157-182.

Murray, J. D., 1975, The structure and petrology of the San José pluton of Baja California, Mexico [Ph.D. thesis]: Pasadena, California Inst. Technology, in prep.

Nishihara, H., 1957, Origin of the "manto" copper deposits in Lower California, Mexico: Econ. Geology, v. 52, p. 944-951. (Translated from Soc. Geol. Mexicana Bol., v. 21, p. 95-110.)

Nordstrom, C. E., 1970, Lusardi Formation—A post-batholithic Cretaceous conglomerate north of San Diego, California: Geol. Soc. America Bull., v. 81, p. 601-606.

Normark, W. R., and Curray, J. R., 1968, Geology and structure of the tip of Baja California, Mexico: Geol. Soc. America Bull., v. 79, p. 1589-1600.

Orcutt, C. R., 1889, Recent and sub-fossil shells of the Colorado Desert: West Am. Scientist, v. 6, p. 92-93.

―――1921, Pleistocene beds of San Quintin Bay, Lower California: West Am. Scientist, v. 19, p. 23-24.

Orme, A. R., 1971, Deformation of marine terraces between Ensenada and El Rosario, Baja California: Geol. Soc. America Abs. with Programs, v. 3, p. 174-175.

Page, B. M., 1970, Sur-Nacimiento fault zone of California—Continental margin tectonics: Geol. Soc. America Bull., v. 81, p. 667-690.

Perrilliat-Montoya, M., 1968, Fauna del Cretacico y del Terciario del norte de Baja California: México Univ. Nac. Autónoma Inst. Geología, Paleontología Mexicana, no. 25, p. 1-36.

Peterson, G. L., 1970a, Distinctions between Cretaceous and Eocene conglomerates in the San Diego area, southwestern California, in Pacific slope geology of northern Baja California and adjacent Alta California: Am. Assoc. Petroleum Geologists and Soc. Econ. Paleontologists and Mineralogists, Pacific Secs., p. 90-98.

―――1970b, Quaternary deformation of the San Diego area, southwestern California, in Pacific slope geology of northern Baja California and adjacent Alta California: Am. Assoc. Petroleum Geologists and Soc. Econ. Paleontologists and Mineralogists, Pacific Secs., p. 120-126.

Peterson, G. L., and Nordstrom, C. E., 1970, Sub-La Jolla unconformity in the vicinity of San Diego, California: Am. Assoc. Petroleum Geologists Bull., v. 54, p. 265-274.

Phillips, R. P., 1964a, Seismic refraction studies in Gulf of California, in van Andel, Tj. H., and Shor, G. G., Jr., eds., Marine geology of the Gulf of California—A symposium: Am. Assoc. Petroleum Geologists Mem. 3, p. 90-121.

―――1964b, Geophysical investigations in the Gulf of California [Dissert.]: San Diego, Univ. California, 250 p. (Univ. Microfilms, Inc., no. 64-9950.)

―――1966, Reconnaissance geology of some of the northwestern islands in the Gulf of California [abs.]: Geol. Soc. America, Cordilleran Sec. Program, p. 59.

―――1967, Seismic explosion results, in Runcorn, S. K., ed., International dictionary of geophysics: Oxford, Pergamon Press, p. 1347-1349.

Phleger, F. B., 1969, A modern evaporite deposit in Mexico: Am. Assoc. Petroleum Geologists Bull., v. 53, p. 824-829.

Phleger, F. B., and Ewing, G. C., 1959, Sedimentology of some Mexican coastal lagoons:

Tulsa Geol. Soc. Digest, v. 27, p. 47-51.
——1962, Sedimentology and oceanography of coastal lagoons in Baja California, Mexico: Geol. Soc. America Bull., v. 73, p. 145-182.
Pichler, H., and Zeil, W., 1969, The Quaternary "andesite" formation in the high Cordilleras of northern Chile: Geol. Rundschau, v. 58, p. 866-903.
Popenoe, W. P., Imlay, R. W., and Murphy, M. A., 1960, Correlation of the Cretaceous formations of the Pacific coast (United States and northwestern Mexico): Geol. Soc. America Bull., v. 71, p. 1491-1540.
Rankama, K., and Sahama, T., 1950, Geochemistry: Chicago, Chicago Univ. Press, 912 p.
Ramos, J. M., 1885, Trabajos ejecutados por la Comisión exploradora de la Baja California, el año de 1884: Ministerio de Fomento Anales 8VO, México, 174 p.
Robinson, P. T., and Elders, W. A., 1971, Late Cenozoic volcanism in the Imperial Valley, California: Geol. Soc. America Abs. with Programs, v. 3, p. 185.
Ross, D. C., 1972, Petrographic and chemical reconnaissance of some granitic and gneissic rocks near the San Andreas fault from Bodega Head to Cajon Pass, California: U.S. Geol. Survey Prof. Paper 698, 92 p.
Rossetter, R., and Gastil, G., 1971, Isla San Luis, a rift volcano in the Gulf of California: Geol. Soc. America Abs. with Programs, v. 3, p. 187-188.
Roy, R. F., 1963, Heat flow measurements in the United States [Dissert.]: Cambridge, Mass., Harvard Univ.
Rusnak, G. A., and Fisher, R. L., 1964, Structural history and evolution of Gulf of California, *in* van Andel, Tj. H., and Shor, G. G., Jr., eds., Marine geology of the Gulf of California—A symposium: Am. Assoc. Petroleum Geologists Mem. 3, p. 144-156.
Rusnak, G. A., Fisher, R. L., and Shepard, F. P., 1964, Bathymetry and faults of Gulf of California, *in* van Andel, Tj. H., and Shor, G. G., Jr., eds., Marine geology of the Gulf of California—A symposium: Am. Assoc. Petroleum Geologists Mem. 3, p. 59-75.
Santillán, M., and Barrera, T., 1930, Las posibilidades petrolíferas en la costa occidental de le Baja California, entre los paralelos 30 y 32 de latitud norte: México Inst. Geología Anales, v. 5, p. 1-37.
Sauer, C. O., 1929, Land forms of the Peninsular Range of California as developed about Warner's Hot Springs and Mesa Grande: California Univ. Pubs. Geography, v. 3, p. 199-290.
Schwarcz, H. P., 1969, Pre-Cretaceous sedimentation and metamorphism in the Winchester area, northern Peninsular Ranges, California: Geol. Soc. America Spec. Paper 100, 63 p.
Sharp, R. V., 1967, San Jacinto fault zone in the Peninsular Ranges of southern California: Geol. Soc. America Bull., v. 78, no. 6, p. 705-729.
Shepard, F. P., 1950, Submarine topography of the Gulf of California, Pt. 3 *of* 1940 *E. W. Scripps* cruise to the Gulf of California: Geol. Soc. America Mem. 43, 32 p.
Shepard, F. P., and Emery, K. O., 1941, Submarine topography off the California coast; canyons and tectonic interpretations: Geol. Soc. America Spec. Paper 31, 171 p.
Shor, G. G., Jr., and Raitt, R. W., 1958, Seismic studies in the southern California continental borderland, *in* Geofísica aplicada, tomo 2: Internat. Geol. Cong., 20th, México [D.F.], 1956 (Trabajos), sec. 9, p. 243-259.
Shor, G. G., Jr., and Roberts, E., 1958, San Miguel, Baja California norte, earthquakes of February 1956—A field report: Seismol. Soc. America Bull., v. 48, p. 101-116.
Silberling, N. J., Schoellhamer, J. E., Gray, C. H., Jr., and Imlay, R. W., 1961, Upper Jurassic fossils from Bedford Canyon Formation, southern California: Am. Assoc. Petroleum Geologists Bull., v. 45, no. 10, p. 1746-1748.
Silver, E. A., 1971, Transitional tectonics and late Cenozoic structure of the continental margin off northernmost California: Geol. Soc. America Bull., v. 82, p. 1-22.
Silver, L. T., 1968, Pre-Cretaceous basement rocks and their bearing on large-scale displacements in the San Andreas fault system [abs.], *in* Conference on geologic problems

of San Andreas fault system, Stanford, Calif., 1967, Proc.: Stanford Univ. Pubs. Geol. Sci., v. 11, p. 279.
Silver, L. T., Stehle, F. G., and Allen, C. R., 1963, Lower Cretaceous prebatholithic rocks of northern Baja California, Mexico: Am. Assoc. Petroleum Geologists Bull., v. 47, p. 2054-2059.
Silver, L. T., Allen, C. R., and Stehle, F. G., 1969, Geological and geochronological observations on a portion of the Peninsular Range batholith of northwestern Baja California, Mexico [abs.]: Geol. Soc. America Spec. Paper 121, p. 279-280.
Sliter, W. V., 1968, Upper Cretaceous foraminifera from southern California and northwestern Baja California, Mexico: Kansas Univ. Paleont. Contr., ser. 49: Protozoa, art. 7, p. 1-141.
Slyker, R. G., Jr., 1969, Geological and geophysical reconnaissance of the Valle de San Felipe region, Baja California, Mexico [abs.]: Geol. Soc. America Spec. Paper 121, p. 560.
Smith, D. L., Roy, R. F., and Smith, A. R., 1971, Distribution of heat producing elements in the Sierra San Pedro Martir, Baja California, Mexico: Geol. Soc. America Abs. with Programs, v. 3, p. 196.
Smith, J. P., 1898, Geographic relations of the Triassic of California: Jour. Geology, v. 6, p. 776-786.
Smith, P. B., 1970, New evidence for a Pliocene marine embayment along the lower Colorado River area, California and Arizona: Geol. Soc. America Bull., v. 81, p. 1411-1420.
Snyder, H., 1970, Mesozoic dike swarms in Baja California: Geol. Soc. America Abs. with Programs, v. 2, p. 145-146.
Sommer, M. A., and Garcia, J., 1970, Potassium-argon dates for Pliocene rhyolite sequences east of Puertocitos, Baja California: Geol. Soc. America Abs. with Programs, v. 2, p. 146.
Spiess, F. N., 1963, El marco geofísico del Gulfo de California: Soc. Geol. Mexicana Bol., v. 26, p. 67-74.
Stewart, H. B., Jr., 1958, Sedimentary reflections of depositional environment in San Miguel Lagoon, Baja California, Mexico: Am. Assoc. Petroleum Geologists Bull., v. 42, p. 2567-2618.
Suppe, J., 1970, Offset of late Mesozoic basement terrains by the San Andreas fault system: Geol. Soc. America Bull., v. 81, p. 3253-3258.
Suppe, J., and Armstrong, R. L., 1972, Potassium-argon dating of Franciscan metamorphic rocks: Am. Jour. Sci., v. 272, p. 217-233.
Sykes, L. R., 1967a, Mechanism of earthquakes and nature of faulting on the mid-oceanic ridges: Jour. Geophys. Research, v. 72, no. 8, p. 2131-2153.
——1967b, Seismicity, in IUGG quadrennial report (U.S.A.): Am. Geophys. Union Trans., v. 48, p. 385-389.
Tamayo, J. L., 1958, Estados Unidos Mexicanos escala 1:100,000: Secretaría de Recursos Hidráulicos.
Tarbet, L. A., 1951, Imperial Valley, California, in Possible future petroleum provinces of North America: Am. Assoc. Petroleum Geologists Bull., v. 35, p. 260-263.
Tarbet, L. A., and Holman, W. H., 1944, Stratigraphy and micropaleontology of the west side of Imperial Valley, California [abs.]: Am. Assoc. Petroleum Geologists Bull., v. 28, p. 1781.
Thatcher, W., 1971, Regional variations of seismic source parameters in the northern Baja California area [abs.]: EOS (Am. Geophys. Union Trans.), v. 52, p. 863.
Thatcher, W., and Brune, J. N., 1969, Surface waves and crustal structure in the Gulf of California, Baja California, and Sonora [abs.]: EOS (Am. Geophys. Union Trans.), v. 50, p. 240.
——1971, Seismic study of an oceanic ridge earthquake swarm in the Gulf of California: Royal Astron. Soc. Geophys. Jour., v. 22, p. 473-482.

Thatcher, W., Brune, J. N., and Clay, D. N., 1971, Seismic evidence on the crustal structure of the Imperial Valley region: Geol. Soc. America Abs. with Programs, v. 3, p. 208.

Thompson, R. W., 1968, Tidal flat sedimentation on the Colorado River delta, northwestern Gulf of California: Geol. Soc. America Mem. 107, 133 p.

Touwaide, M. E., 1930, Origin of the Boleo copper deposit, Lower California, Mexico: Econ. Geology, v. 25, p. 113-144.

Turner, F. J., and Verhoogen, J., 1960, Igneous and metamorphic petrology (2d ed.): New York, McGraw-Hill Book Co., 694 p.

Uchupi, E., and Emery, K. O., 1963, The continental slope between San Francisco, California and Cedros Island, Mexico: Deep-Sea Research, v. 10, p. 397-447.

U.S. Army Map Service, 1958-1964, Estados Unidos Mexicanos: U.S. Army Corps of Engineers, Series F 501, ed. 1, scale 1:250,000.

Valentine, J. W., and Rowland, R. W., 1969, Pleistocene invertebrates from northwestern Baja California del Norte, Mexico: California Acad. Sci. Proc., ser. 4, v. 36, p. 511-530.

van Andel, Tj. H., 1963, Algunos aspectos de la sedimentación reciente en el Golfo de California: Soc. Geol. Mexicana Bol., v. 26, no. 26, p. 85-94.

——1964, Recent marine sediments of Gulf of California, in van Andel, Tj. H., and Shor, G. G., Jr., eds., Marine geology of the Gulf of California—A symposium: Am. Assoc. Petroleum Geologists Mem. 3, p. 216-310.

van Andel, Tj. H., and Shor, G. G., Jr., eds., 1964, Marine geology of the Gulf of California—A symposium: Am. Assoc. Petroleum Geologists Mem. 3, 408 p.

Vaughn, T. W., 1900, A new fossil species of *Caryophyllia* from California: U.S. Natl. Mus. Proc. 22, p. 199-203.

——1917, The reef-coral fauna of Corrizo Creek, Imperial County, California, and its significance: U.S. Geol. Survey Prof. Paper 98, p. 355-395.

Velasco, H. J., 1970, Levantamiento gravimétrico zona geotérmico de Mexicali, Baja California: México Consejo Recursos Nat. no Renovables Bol. 74, 19 p.

von Chrustschoff, K., 1885, Ueber ein neues aussereuropaisches Leucitgestein: Tschermaks Mineralog. u. Petrog. Mitt., ser. 2, v. 6, p. 160-171.

vonEngeln, O. D., 1942, Geomorphology: New York, Macmillan Co., 655 p.

Von Herzen, R. P., 1963, Geothermal heat flow in the gulfs of California and Aden: Science, v. 140, p. 1207-1208.

——1964, Ocean-floor heat-flow measurements west of the United States and Baja California: Marine Geology, v. 1, p. 225-239.

Von Herzen, R. P., and Maxwell, A. E., 1964, Measurements of heat flow at the preliminary Mohole site off Mexico: Jour. Geophys. Research, v. 69, p. 741-748.

Walker, T. R., 1967, Formation of red beds in modern and ancient deserts: Geol. Soc. America Bull., v. 78, p. 353-368.

Walker, T. R., and Thompson, R. W., 1968, Late Quaternary geology of the San Felipe area, Baja California, Mexico: Jour. Geology, v. 76, p. 479-485.

Weaver, D. W., 1969, The pre-Tertiary rocks, in Geology of the Northern Channel Islands, southern California borderland: Am. Assoc. Petroleum Geologists and Soc. Econ. Paleontologists and Mineralogists, Pacific Secs., p. 11-13.

Webb, R. W., 1939, Evidence of the age of the crystalline limestone in southern California: Jour. Geology, v. 47, p. 198-201.

Weber, F. H., Jr., 1963, Geology and mineral resources of San Diego County, California: California Div. Mines and Geology, County Rept. 3, 309 p.

Wegener, A., 1928, Die Entstehung der Kontinente and Ozeane (English transl.): London, T. Murby & Co., 240 p.

White, C. A., 1885, On new Cretaceous fossils from California: U.S. Geol. Survey Bull. 22, p. 355-373.

White, D. E., Anderson, E. T., and Grubb, D. K., 1963, Geothermal brine well—Mile-deep hole may tap ore-bearing magmatic water and rocks undergoing metamorphism: Science,

v. 139, p. 1919-1922.

Wiegand, J. W., 1970, Evidence of a San Diego-Tijuana fault: Assoc. Eng. Geologists Bull., v. 7, p. 107-121.

Willis, B., and Stose, G. W., 1912, Index to the stratigraphy of North America: U.S. Geol. Survey Prof. Paper 71, 896 p.

Wilson, I. F., 1948, Buried topography, initial structure, and sedimentation in Santa Rosalía area, Baja California, Mexico: Am. Assoc. Petroleum Geologists Bull., v. 32, p. 1762-1807.

Wilson, I. F., and Rocha Moreno, V. S., 1955, Geology and mineral deposits of the Boleo copper district, Baja California, Mexico: U.S. Geol. Survey Prof. Paper 273, 134 p.

Wilson, J. T., 1965, A new class of faults and their bearing on continental drift: Nature, v. 207, no. 4995, p. 343-347.

Wisser, E., 1954, Geology and ore deposits of Baja California, Mexico: Econ. Geology, v. 49, p. 44-76.

Wittich, E., 1909, Contribuciones á la geología de la región meridional de la Baja California: Soc. Geol. Mexicana Bol., v. 6, p. 5-14.

——1911, Beitrage zur Geologie der Kapregion von Nieder-Kalifornien: Deutsch. Geol. Gesell. Zeitschr., Monatsber., v. 63, p. 578-587.

——1915, Über Eisenerzlager an der Nordwestküste von Nieder-Kalifornien: Zentralblatt für Mineralogie, Geologie und Paläontologie, v. 16, p. 389-395.

Wood, H. O., and Heck, N. A., 1951, Stronger earthquakes of California and western Nevada, in Earthquake history of the United States, Pt. 2: U.S. Coast and Geod. Survey [Pub.], ser. 609 (revised), 35 p.

Woodard, G. D., 1974, Redefinition of Cenozoic stratigraphic column in Split Mountain gorge, Imperial Valley, California: Am. Assoc. Petroleum Geologists Bull., v. 58, p. 521-526.

Woodford, A. O., 1925, The San Onofre Breccia; its nature and origin: California Univ., Dept. Geology Bull., v. 15, no. 7, p. 159-280.

——1928, The San Quintin volcanic field, Lower California: Am. Jour. Sci., ser. 5, v. 15, p. 337-345.

Woodford, A. O., and Harriss, T. F., 1938, Geological reconnaissance across Sierra San Pedro Mártir, Baja California: Geol. Soc. America Bull., v. 49, p. 1297-1336.

Woodford, A. O., Welday, E. E., and Merriam, R., 1968, Siliceous tuff clasts in the upper Paleogene of southern California: Geol. Soc. America Bull., v. 79, p. 1461-1486.

Woodring, W. P., 1931, Distribution and age of the marine Tertiary deposits of the Colorado Desert: Carnegie Inst. Washington Pub. 418, Contr. Paleontology, p. 1-25.

Wright, L. B., 1946, Geology of Santa Rosa Mountain area, Riverside County, California: California Jour. Mines and Geology, v. 42, p. 9-13.

Yeats, R. S., Minch, J. A., and Forman, J. A., 1971, Paired basement terranes in Baja California Sur, Mexico: Geol. Soc. America Abs. with Programs, v. 3, p. 760.

Yerkes, R. F., 1957, Volcanic rocks of the El Modeno area, Orange County, California: U.S. Geol. Survey Prof. Paper 274-L, p. 313-334.

Zárate, D., 1922, A trip to the mining district of Real del Castillo, Agua Dulce, Jacalitos, Guatay and El Burro: El Hispano Americano, Lower California ed., p. 48.

Zárate, D., and Nuñez, J., 1925, Ensenada y su origen: El Hispano Americano (San Diego), Sept. 16.

MANUSCRIPT RECEIVED BY THE SOCIETY JUNE 15, 1971
REVISED MANUSCRIPT RECEIVED NOVEMBER 2, 1973